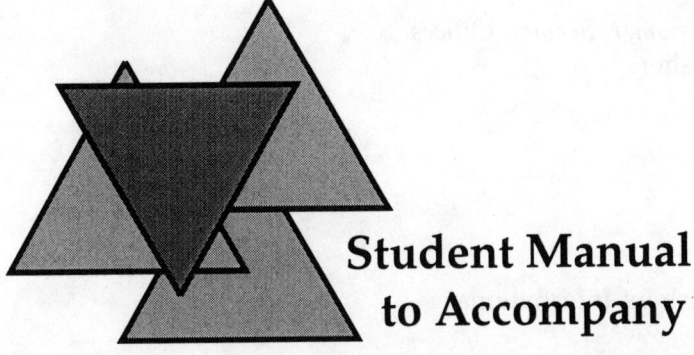

Student Manual to Accompany

Essentials of Molecular Biology

Third Edition

A User-Friendly Guide for Learning Molecular Biology

George M. Malacinski
Pamela L. Hanratty
Indiana University

JONES AND BARTLETT PUBLISHERS
Sudbury, Massachusetts
BOSTON LONDON SINGAPORE

Editorial, Sales, and Customer Service Offices
Jones and Bartlett Publishers
40 Tall Pine Drive
Sudbury, MA 01776
978-443-5000
info@jbpub.com
http://www.jbpub.com

Jones and Bartlett Publishers International
Barb House, Barb Mews
London W6 7PA
UK

ISBN: 0-7637-0417-2

Printed in the United States of America
02 01 00 99 98 97 5 4 3 2 1

Contents

INTRODUCTION

Foreword

Essentials of Molecular Biology, Third Edition, is designed as an entry-level textbook for sophomore/junior undergraduate courses. A "layering" approach to the discipline of molecular biology is taken, whereby complexity is developed in a sequential fashion. Concepts, ideas, and models of molecular mechanisms are emphasized. Students are guided by the textbook toward an understanding of each issue, as well as toward an appreciation for the significance of concepts and technologies encompassed by the discipline of molecular biology.

To insure that students learn the "essential" features of molecular biology, without being distracted by excess information, the second edition of *Essentials of Molecular Biology* adopted the following motto:

It is better to understand a little, than to misunderstand a lot.

An "anti-encyclopedic" approach to learning molecular biology is employed by *Essentials of Molecular Biology , Third Edition,* and is reinforced by this Student Manual. In contrast to the encyclopedic texts (the Appendix of this manual lists several of them), *Essentials of Molecular Biology , Third Edition,* focuses on explaining only the key observations, ideas, facts, and concepts that comprise the repertoire of molecular mechanisms thus far known to be associated with gene expression and cell function. The use of excess data (e.g., tables), complex illustrations, and tangential examples are assiduously avoided. The more complicated information and descriptions included in *Essentials of Molecular Biology , Third Edition,* are, in many instances, amplified in this Student Manual. A somewhat unique style has been adopted to insure that students (and instructors) find this study manual useful.

Style and Format

The design of this study manual was developed largely by undergraduate students who completed the Indiana University sophomore-level molecular biology course that uses *Essentials of Molecular Biology , Third Edition,* as the textbook. Accordingly, emphasis has been placed on preparing a Student Manual that emphasizes "understanding" and depreciates the value of "rote memorization." In fact, the following logos have been adopted in our molecular biology course:

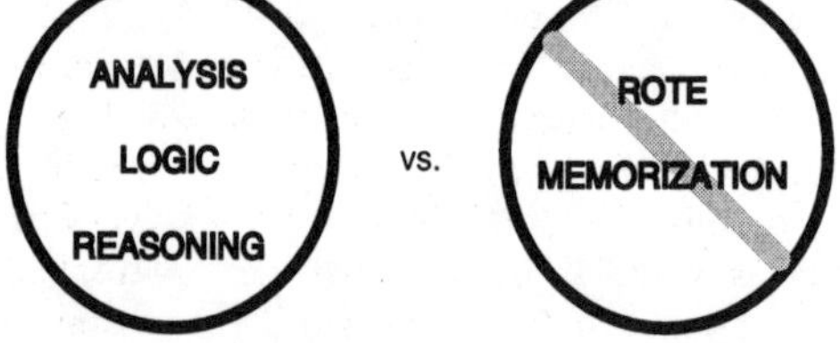

Three key stylistic features have been employed throughout this Student Manual:

Conversational prose has been adopted as one of the dominant stylistic themes. This is to insure that the Student Manual is "user friendly" and not so serious as to be intimidating to beginning molecular biology students. A deliberate attempt has been made to avoid "molecular biology jargon." From one of the encyclopedic texts listed in the Appendix, the following short paragraph provides an example of molecular biology jargon.

"Two factors intrinsic to the double helix determine the likelihood that a particular region of DNA will exist in the Z-form. One is the nucleotide sequence, the other is the overall structure of the double helix in the sense of its path in space, an effect described in the next section. If both of these factors are propitious, DNA may be able to convert from B-form to Z-form in the natural state."

That sort of prose is avoided at all costs in this Student Manual!

Interactive explanation processes have been used whenever appropriate. In some instances, an explanation has been followed by an inquiry. In other cases, "do-it-yourself" exercises have been employed to aid students in understanding the material and facilitating knowledge construction.

Relevance of the discipline of molecular biology to practical matters (e.g., health care issues) and societal issues (e.g., potential ethical conflicts) have also been explained wherever appropriate.

These features distinguish this study guide and its cognate textbook, *Essentials of Molecular Biology, Third Edition,* as being distinct from the encyclopedic texts described in the Appendix.

How to get started

Students should figure out their individual learning styles!

Mr. Paul W. Zell, a project development specialist formerly associated with teaching Molecular Biology at Indiana University, has prepared the following descriptions of learning styles. Please read through them and "figure yourself out"!

A Guide to Learning Styles for the Beginning Student of Molecular Biology

Paul W. Zell • *Indiana University*

You live in a great time to be a college student. The center of attention in higher education in past years focused almost entirely on the presentation of the teacher. More recently, that focus has been shifted so that now the largest concern is the capability and skill of the student. The third edition of the textbook *Essentials of Molecular Biology* has been carefully edited with the central focus on the learner. Likewise, this Student Manual has been designed to be user-friendly.

As with most of the real world, where there is added privilege, there is also added responsibility. Our expectation is that you will respond positively to the challenge of learning molecular biology from *Essentials of Molecular Biology, Third Edition.* The purpose of this introduction to learning styles is to help you take control of your own learning and increase your level of performance.

Isn't it funny how most of us want to be recognized as individuals; yet at the same time, we don't want to be seen as too different from our peer group? Comparing our personalities with those of other people reveals differences and similarities. In this guide we will look at how these differences and similarities occur in typical patterns. The profiles presented here characterize (not categorize) the similarities and differences in what we now call *learning style.*

Styles vary across a common process. Each style has its advantage. No style is inherently superior to any other. Each style is "custom fit" to be superior for that individual. **The same is true for learning styles.** In Molecular Biology, different topics play to the strengths of different learning styles, but all topics can be mastered regardless of your individual learning style. The profiles characterized here are analogous to dexterity. Most of us are more skillful in the use of one hand over the other. But that doesn't mean we are restricted to using only one hand. With that in mind, let's consider some of the differences in learning styles.

Everyone uses at least four basic mental processes: Sensing, Intuition, Thinking, and Feeling. Both Sensing and Intuition are types of perception. You probably prefer one over the other. They describe your preference for becoming aware of things, people, events, and ideas. Sensing perception is possessed by those who are observant, realistic, and tend to focus on immediate experiences available to the five senses. Their characteristics include enjoyment of the present, keen observational powers, and a memory for detail. Sensing students take detailed lecture notes and prefer classes where hands-on work and concrete experience are emphasized. They will easily and quickly master the chemical formulas included in Chapter 2 of *Essentials of Molecular Biology, Third Edition.*

Those who favor Intuitive perception tend to perceive holistically. Possible meanings and relationships may suddenly come to the surface from material that may seem unrelated to other people. When Sensing students approach a problem, they look for details. When Intuition students attack a problem, they may overlook some details but will offer new possibilities for viewing the problem, and, hence, provide a framework for holding the details together. Problems are stimulating rather than deflating. Chapters 11 and 13 of *Essentials of Molecular Biology, Third Edition,* on gene regulation will appeal to Intuitive students.

The rational judgment processes are Thinking and Feeling. Thinking students tend to be analytical and impersonal. Thinking students are motivated by logical rationale. Their desire is to understand cause-and-effect relationships. However, they may become frustrated when approached with an interpersonal or social problem and find that their favorite learning style is not well-suited to this type of problem.

Feeling mode learning is accomplished by weighing the relative values and merits of issues. (Note that the emphasis is on values, and not emotions.) Feeling students prefer topics that have a human angle. They want to have their personal convictions and values considered. Since Feeling relies on understanding the values of other people, it is more subjective than Thinking. Therefore, it is very useful for solving problems of an interpersonal nature.

There is also a noteworthy difference in our attitudes toward things we encounter. We may be Extroverted or Introverted. Extroversion describes an attitude that pays attention to the external world of objects and people. Extroverts actively try to interact with people in learning situations. They think best when talking and they learn well in groups. Because they value active experience, they often use a strategy of trial and error since it allows them to think while acting.

Introversion describes an attitude that directs energy toward the internal world of concepts and ideas. (Note that Introvert does not necessarily mean "shy.") Introverts think best when alone and need quiet time for concentration and study. Teacher-controlled and lecture-based classes are geared more toward Introverts than toward Extroverts. Because they do not always share this internal world with others, they may not perform especially well in class discussions, and teachers may be slow to appreciate their grasp of the subject.

The final difference we will consider describes student preferences for structured learning environments. These two preferences are called Judgment and Perception. Judging types prefer structured situations that provide goals and deadlines. They like to gather some information and then come to a reasoned decision. Perceptive types prefer a more flexible approach, enabling them to thoroughly analyze the problem at hand. Perceptive types are sometimes viewed as procrastinators because they gather as much information and data as possible before starting or finishing a project.

LEARNING STYLE and READING THE TEXT

The active learning preferred by Extroverts may make it difficult for them to spend long periods of time in concentrated reading. They might find Chapter 7 in *Essentials of Molecular Biology, Third Edition,* on DNA replication a bit tedious. Extroverts sometimes understand a text better if they read aloud. If you find a section that is particularly difficult, try reading it out loud. Or, find a study partner and discuss the chapter paragraph by paragraph. Although Introverts have less difficulty concentrating on the text, they sometimes lose concentration because they get lost in contemplation. They should use the study questions at the end of each chapter of *Essentials of Molecular Biology, Third Edition,* to guide their contemplation. This Student Manual can also be used to keep focused on the subjects discussed in the main text.

Sensing students often focus on the facts in a textbook and neglect the larger concepts. As a result, they have difficulty seeing how the facts fit together. They also tend to become bogged down in facts. Intuitive learners, on the other hand, may focus on the concepts and neglect the facts. They tend to miss the individual details that give reasoned support for the concept.

The textbook *Essentials of Molecular Biology , Third Edition*, and this Student Manual try to engage both Thinkers and Feelers. Thinkers focus on the message and neglect the style. They probably have more love for theory and abstraction than Feelers. However, Feeling students often become bored with books that don't present science as a human endeavor reflecting personal values. The field of molecular biology needs both types of learners!

Judging students need to avoid being too quick to interpret a text with a cursory reading. Perceiving students need to a put a limit on their investigation and avoid becoming bogged down in the reading.

Clearly, no single learning style should dominate a beginning Molecular Biology class. You will find some chapters to be your niche. Use and build on the strengths you have. But be prepared to adapt your style in other places. Always keep developing as a person. Above all, enjoy your journey through *Essentials of Molecular Biology, Third Edition* !

The Real World

What good is a biology undergraduate degree, anyway?

Please keep in mind that biology, especially molecular biology, provides a training ground for intellectual development. There are literally hundreds of types of jobs available for biologists. The problem-solving skills that students of *Essentials of Molecular Biology, Third Edition,* learn provide a stepping stone for thousands of employment possibilities.

How does a background in Molecular Biology benefit a "pre-med" student?

In the short term, a lot. New versions of the Medical College Admissions Test (MCAT) include fewer factual questions and more questions that assess a student's capacity for analytical thinking. Likewise, *Essentials of Molecular Biology, Third Edition,* keeps the facts to a minimum, and this Student Manual emphasizes problem-solving skills, analysis, and understanding.

In the long term, even more. Molecular medicine represents a discipline that is growing at a phenomenal rate. New insights into disease phenomena, designer drugs, and gene therapy frequent newspaper and journal headlines. The discipline of molecular biology has led to a vast number of recent advances in medicine!

Should I go on to graduate school for a Ph.D.?

There exists no single, general answer to this question. The answer must be custom-tailored to meet the needs of each individual. You can use this manual to start developing your own answer.

From one college professor comes the following quote:

> "Creative people have a positive attitude toward problem solving. They consider a problem to be a challenge, an opportunity for new experiences, an enrichment of the repertoire of tools for thinking, and a learning experience. With a positive attitude, a frustrated effort to identify a solution is deemed to be compensated for in great measure by the lessons that can be learned when no solution is found. Creative people view an obstacle in a problem-solving situation as a challenge—an intellectual and emotional adventure. Creative people do not run away from complex situations. They tolerate complexity, uncertainty, conflict, and dissonance. They enjoy new experiences. They are more active than passive, and they have capacity for producing results. They are doers. They seem to be in control. They radiate self-confidence."

First: Does this quote make sense to you? Does it describe a lot of yourself? Would you like to be like the kind of person described therein?

Second: Review the biosketches of the *Essentials of Molecular Biology, Third Edition,* chapter editors. Do you see role models among these scientists? That is, do their backgrounds, interests, and personal statements make you feel comfortable?

Third: Assess your potential for academic development. Does learning fascinate you or do you find it boring? Do you routinely achieve good and/or "top" grades in your courses? Do you enjoy and do well in problem-solving and analytical courses as well as descriptive ones? Do your personal circumstances and life style provide you with the opportunity to commit yourself to 4-6 years of postbaccalaureate education?

Fourth: Read below what several graduate students (at Indiana University) had to say about "why I chose to attend graduate school."

> "To obtain the qualifications to direct an independent research project."
>
> "I didn't want to be a part of industry, and only academia provides an opportunity to pursue my interests (molecular evolution)."
>
> "To pursue a career in teaching."
>
> "After several years of working as a technical assistant, I wanted to advance to the next level."

Fifth: If all the signs point toward graduate school, develop a "game plan." Here is a suggested step-by-step procedure:

1. Talk with one or two of your professors. Most likely they have attended graduate school and can offer you some insights.

2. Take the Graduate Record Examination (GRE). It will be required for processing your applications for graduate study. Your professors can suggest the best semester to do this. Usually, one semester before filing graduate school applications (i.e., the first semester of your senior year) is an appropriate time.

3. Review your track record (course grades, work experience, GRE scores, and the like.) with your professors, and, if possible, with several graduate students. Compare these "quantitative" features of your assessment with your "qualitative" assessment developed above in the third section. Are they compatible? If "yes," proceed to (4) below. If a discrepancy arises, please consult your professor. Talk it over. Perhaps you are a "late bloomer," a "diamond-in-the-rough," or somebody with as-yet unrecognized, or untapped potential.

4. Write letters of inquiry to graduate schools to request an application and descriptive brochure. Address these letters to either the graduate school or the specific department (e.g., biology, molecular biology, genetics, or chemistry). You will likely be pleasantly surprised with the information that you receive:
   a. A lot of opportunities exist.
   b. Financial support, often including tuition, fees, and living allowance, is usually available. That is, most modern graduate departments in science offer *all* entering students financial support.

5. Select several universities in consultation with your professors and various graduate students.

6. Apply!

As you develop a commitment to attend graduate school, please keep an open mind about career options. Since your main, and perhaps only, exposure to professional biologists has been with your professors, you might naturally imagine yourself also becoming a professor because you admire their intellectualism and respect their lifestyles. However, nonacademic employment possibilities for Ph.D.s are ever expanding and will engage an increasing proportion of biology Ph.D.s in the next few decades. The biotechnology industry, government forensic medicine laboratories, and health professions (e.g., genetic counseling) are all expanding and expect to hire more biology Ph.D.s than ever before!

If several graduate schools accept me, how should I choose which one to attend?

Here are what several Indiana University graduate students said in reply to the following question: "Why did you choose I.U.?" Perhaps their answers can help direct your attention to some important factors that merit consideration.

"I like the graduate program."

"The research interests of the faculty appealed to me."

"Graduate students impressed me as being happy."

"The financial-aid package was good."

Acknowledgments

Over the course of a four-year period approximately 30 undergraduate students contributed suggestions, editorial comments, and modifications to this Student Manual. They are hereby acknowledged for their contributions.

Special Thanks

Cregg Ashcraft provided many of the entries in "Here's Help" and generated many of the study questions and answers. He also offered valuable assistance in other phases of this project.

Sharon Achilles contributed expert editorial assistance. She substantially improved this manual by carefully reworking the prose and format.

Randy Van Horn supervised the preparation of some of the study questions and answers. He also organized the writing of weekly worksheets that are essential to the collaborative learning approach that is taken in teaching molecular biology at Indiana University (see Appendix B.). A sample worksheet is included in the Instructor's Guide section of the Appendix.

Note to students

Learning is work. Usually it is *"hard"* work. This Student Manual does not pretend otherwise. It does, however, use a stylistic approach that tries to make the work associated with learning molecular biology *"fun."* Enjoy!

1

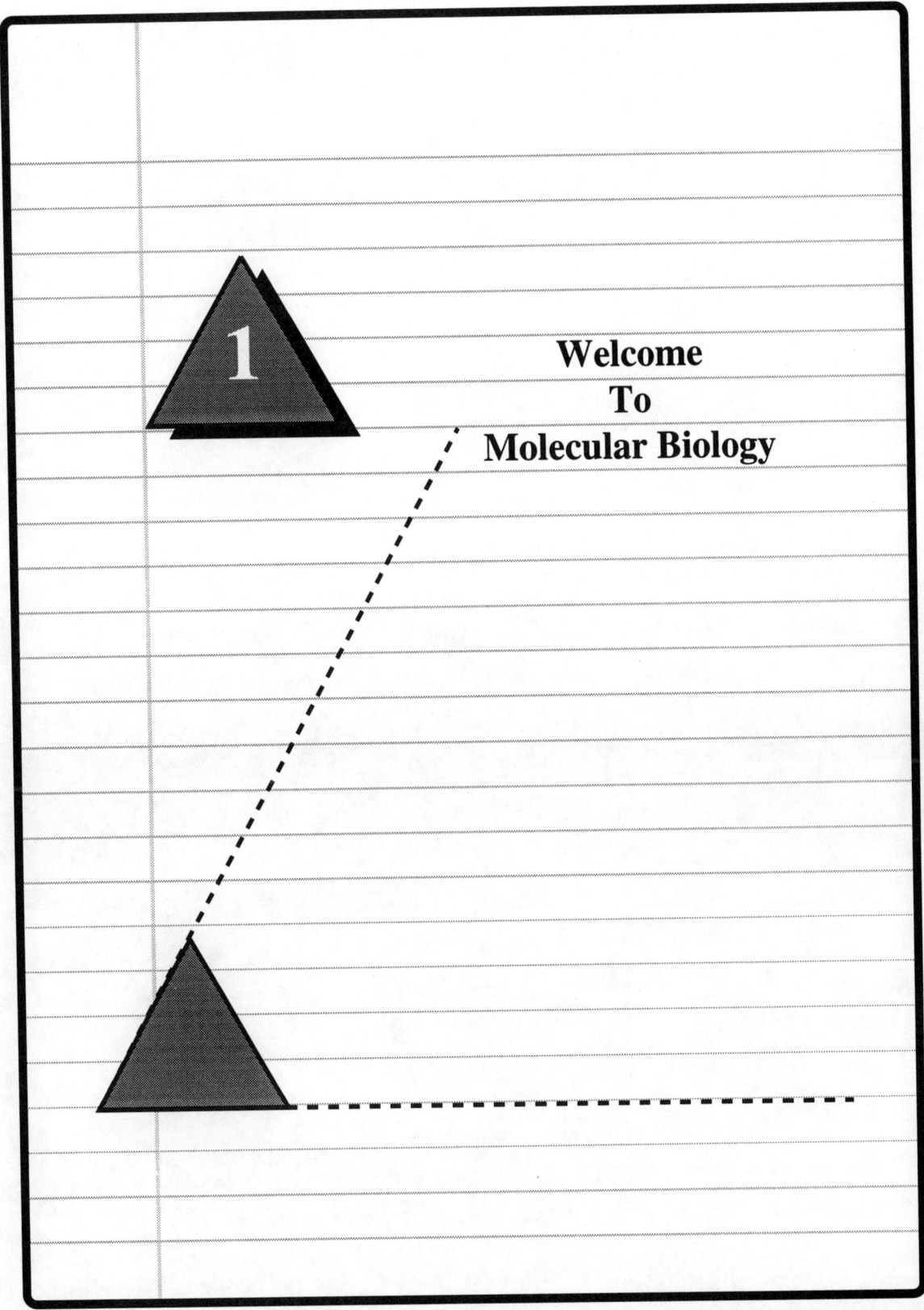

Welcome To Molecular Biology

Here's Help

What factors propelled progress in Molecular Biology?

Several factors came into play over the course of the 20-30 years of the early history of Molecular Biology. Included among those factors was the development of a variety of experimental methods for the characterization of the physical and chemical features of the polymers (RNA, DNA and protein) which comprise the information flow pathways in living cells (see Table 1-2). As those methods were developed a number of physicists and biochemists began to explore simple living systems, especially viruses and their bacterial hosts, in search of a set of basic principles which govern the behavior of living cells. As they began to succeed, Molecular Genetics as a discipline emerged. Simple organisms were employed for learning about the structure of genes and the manner in which they are expressed.

As progress snowballed, more and more challenges were undertaken by the increasingly successful practitioners of Molecular Biology. Eventually, an especially bold initiative—sequencing of the entire human genome (so called "Human Genome Project")—was begun.

What factors limit progress in Molecular Biology?

Many scientists will say that for Molecular Biology the "sky is the limit" (see a brief report in Science, vol. 236, p. 1518 [1987]) and that only a shortage of research funding will slow the rate of progress. Others argue that Molecular Biology as a discipline has two inherent weaknesses: (1) It lacks predictive power; and (2) The behavior of the whole living cell is much more complex than a simple summation of its individual (molecular) parts. (See a short essay in BioEssays, vol. 3, p. 3 [1985]).

For many projects, including the discovery of many disease-causing genes, the former appears certainly to be true. For other projects, including the analysis of human memory, the latter view which emphasizes "limitations" appears to be especially relevant.

For yet other projects, such as the study of "human consciousness" Molecular Biology is not yet even poised to make an assault. Nevertheless, just reading newspaper reports should suffice to verify the remarkable progress that is being made by today's molecular biologists!

Why use yeast as an experimental organism?

There are several reasons why yeast make good experimental organisms. They are small, unicellular, and they grow and reproduce quickly. We will focus on two additional reasons for using yeast: they are eukaryotes, and they can exist in either haploid or diploid states. Two species of yeast, *Saccharomyces cerevisiae* and *Schizosaccharomyces pombe*, are commonly used in the laboratory. However, we will always refer to the former in this text.

Bacteria, like yeast, are small unicellular organisms that reproduce quickly and are easy to maintain and manipulate in a laboratory situation. Bacteria, however, are prokaryotes. Many problems arise when applying data gained from a prokaryotic system to a eukaryotic system. One advantage to using yeast, although a very simple system, is that they are at least eukaryotes and can therefore be used to make hypotheses about higher organisms.

During its life cycle, a yeast may exist in either a haploid or a diploid state. When in the haploid state, *S. cerevisiae* exists as one of two mating types, either "a" or "A." Under certain conditions, haploid yeast cells of opposite mating types "conjugate" to combine their haploid genomes and form a single diploid cell. This diploid cell can either divide mitotically, to form two diploid daughter cells, or it can divide meiotically, to form four haploid daughter cells. This is a tremendous asset to genetic analyses, especially when performing complementation tests. To perform these tests using bacteria, one must insert DNA into the bacterial cell in order to form a partial diploid, but when using yeast, all one has to do is mate two haploid cells.

What advantages do "primary" mammalian cell cultures offer over "tumor cells in culture" (which can be cultured over a longer period of time)?

Primary cell cultures are prepared from fresh tissue, which is usually minced and placed in a rich culture broth. The cells in such a culture broth usually continue to function in much the same manner in which they functioned while in a tissue (e.g., muscle cells can be stimulated to contract). Thus, primary cell cultures are very advantageous for many experimental designs. However, most tissues consist of mixed populations of cells. Muscle tissue, for example, although consisting predominantly of contractile cells, also contains connective tissue cells. So, if a "pure" culture of cells is required, primary cell cultures are not satisfactory. In addition, primary cell cultures often die out relatively quickly (e.g., two months after first culturing).

Tumor cells, while favorable because they can be cultured as extensions of a single starting culture for several years, quickly lose the distinguishing features of the tissue from which they were initially obtained.

Do the so-called "logics" of Molecular Biology apply to other academic disciplines?

Yes, for sure. It is, however, a matter of emphasis. Because molecular biology is so soundly based in "experimental" procedures, its thought processes are somewhat narrowly focused around the scientific method. Heavy reliance on model building and quantitative assessment is common. Other disciplines, such as Anthropology, employ similar logics. But because of Anthropology's heavy dependence upon "retrospective" analyses (vs. direct experimentation), its collection of thought processes tends to emphasize phylogenetic history and strong inference.

Is it "fair" to condense a discipline such as Molecular Biology into a simple set of a couple of dozen concepts?

Yes, and no. Yes, because the inherently reductionist nature of the discipline makes it easy to get lost in minute details and lose sight of the "big picture". Thus, a statement of concepts or principles or themes serves to integrate the details for the beginning student of this discipline.

No, because the features of biological macromolecules evolved in apparently random fashion. Changes in a protein or nucleic acid which improved the fitness of the organism have been perpetuated through generations of natural selection. It therefore can be argued that it is not valid to condense diverse products of widespread random events into a set of concise statements.

However, Essentials of Molecular Biology, designed to be user friendly, votes "yes". That is because beginning Molecular Biology students have requested a set of "guiding principles" as a counterbalance to the reductionist theme of Molecular Biology.

Key Terms

- Auxotroph
- Prototroph
- Minimal medium
- Plaque
- Mutant
- Arabidopsis

Logics of Molecular Biology

- Efficiency argument
- Phylogenetic history
- Quantitative assessment
- Model building
- Parallelism
- Strong inference
- Optimism

For More Information . . .

Where can I easily get some other explanations on these topics?

Information about the bacteriophage life cycle can be found in Alberts, *et. al.*, *Molecular Biology of the Cell*, 3rd edition: p. 281.

The yeast life cycle is explained quite well on page 880 of *Molecular Biology of the Cell.*

The Wise Owl Says

In science, it is usually easier to DISPROVE something than to prove something!

P.S.: A case in point is DNA polymerase I. For years, it was thought to represent the cell's <u>key</u> DNA replication enzyme. Then a ($PolA^-$) mutant was discovered that disproved this notion!

Tough Nuts

I. Although it is relatively easy to perform genetic experiments on bacteriophage, bacteria, and yeast, it is difficult to carry out genetic analyses on vertebrates, especially higher vertebrates with long generation times. Humans, of course, fit in the latter category.

Information, concerning the function of specific macromolecules and the metabolic regulation associated with various cellular processes, is now desired for vertebrate cells. Genetic analysis proved to be extremely useful in obtaining these types of data for bacteria and may, therefore, also be useful for vertebrates.

How can appropriate genetic strategies for vertebrates be designed?

Please consider the following points:

a. Pedigree analyses are often helpful, but they have inherent limitations. What are these limitations?

b. Breeding experiments can be performed for small mammals, especially if the generation time (egg-to-egg life cycle) is relatively short (e.g., 3-6 months). However, what are some of the practical considerations?

c. Isolated mammalian cells are occasionally employed for studies on metabolic regulation. That is, mammalian cells are maintained in culture just as bacterial cells are grown in culture. Do these cells represent a useful model system for the whole animal?

d. "Transgenic" animals can be produced by injecting a purified, foreign gene into the egg of an unrelated species. The gene will occasionally be integrated into the host's chromosomes and, in some instances, will actually be expressed. Can this experimental approach be applied to agriculturally important animals? To humans?

II. Despite rapid progress, inherent limitations in the field of Molecular Biology have been argued. One molecular biologist, Adam S. Wilkens, discussed two of these limitations in an essay a few years ago:*

1. Molecular Biology as a theory is incomplete. Frequently, it offers no theoretical predictions and, therefore, cannot serve as a source of testable hypotheses.
2. No general rules can be formulated to deduce the properties, or behavior, of cells from an inventory of their macromolecules.

How can these limitations be circumvented, or at least accommodated?

* See *Bioessays*, July 1985: The Limits of Molecular Biology. Similar points of view have been expressed by others. Please see:

Tennent, N.W. (1986). Reductionism and Holism in Biology. In *History of Embryology*, eds. T.J. Horder, J.A. Witkowski, and C.C. Wylie, pp. 407-433. Cambridge University Press.

Bard, Jonathan. (1990). *Morphogenesis: The Cellular and Molecular Processes of Developmental Anatomy*. Cambridge University Press, pp. 117-119.

Concept Map

For each chapter of *EMB* we suggest that you organize the knowledge you have gained into a "concept map." As you connect the various terms, experiments, and phenomena presented in the chapters, your leaning experience will be enhanced. The idea is to increase your understanding of the "big picture."

The relationships between the various sections of each chapter will become more apparent to you as you tie them together with arrows. For each map, you will be given a jump start. Several key terms will be included in boxes. A list of additional terms will also be included. As you think about the relationships between individual entries, your understanding of each chapter will be increased.

No single map is correct. In fact, the final format of a map is itself not nearly as important as is the process of preparing the map. The analysis, evaluation, and synthesis exercise involved in drawing the map provides the key learning experience. Since individual students have highly personal conceptual frameworks and different principles have somewhat different meanings to each of us, no two maps will look exactly alike. That is, the position on the hierarchy of a particular concept or term will probably vary somewhat from one student to another.

Once your map is complete, engage in a cooperative learning experience. Compare your map with those of a few classmates to see how they differ. Explain to them the rationale behind the unique features of your map. Through this process, you will undoubtedly enhance your verbal communication skills.

As a final exercise, run down the list of "key terms" found in the chapter and check to be certain you know where they fit into your concept map.

Shall we pause now to tie the information provided in Chapter 1 of *EMB* together into a concept map*?

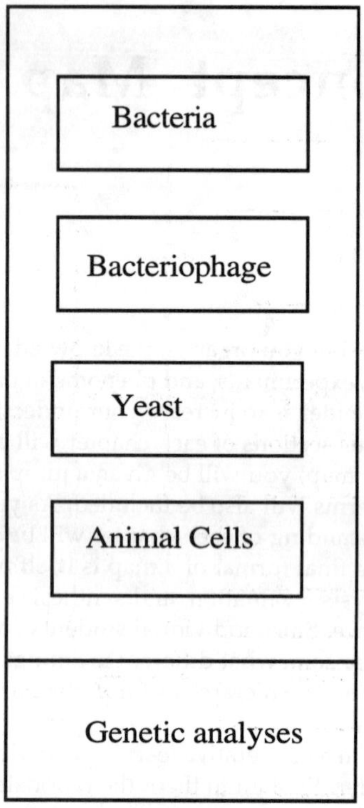

Please add arrows or connecting lines that illustrate logical connections between the above boxes, also plug in with connecting lines or arrows, the following components:

prototroph or auxotroph
culture media
plaque assay
gene function
simple eukaryotic model
metabolic regulation
established cell lines
logic of molecular biology
efficiency argument
quantitative assessment
model building
parallelism
strong inference
phylogenetic history
transgenic animals

*For this chapter, one possible configuration of this concept map is provided on the next page. Keep in mind, however, that an almost limitless variety of equally valid maps could be drawn. Your map should help you learn. Your classmates learning styles and learning techniques are probably somewhat different from yours, so their maps will also be somewhat different!

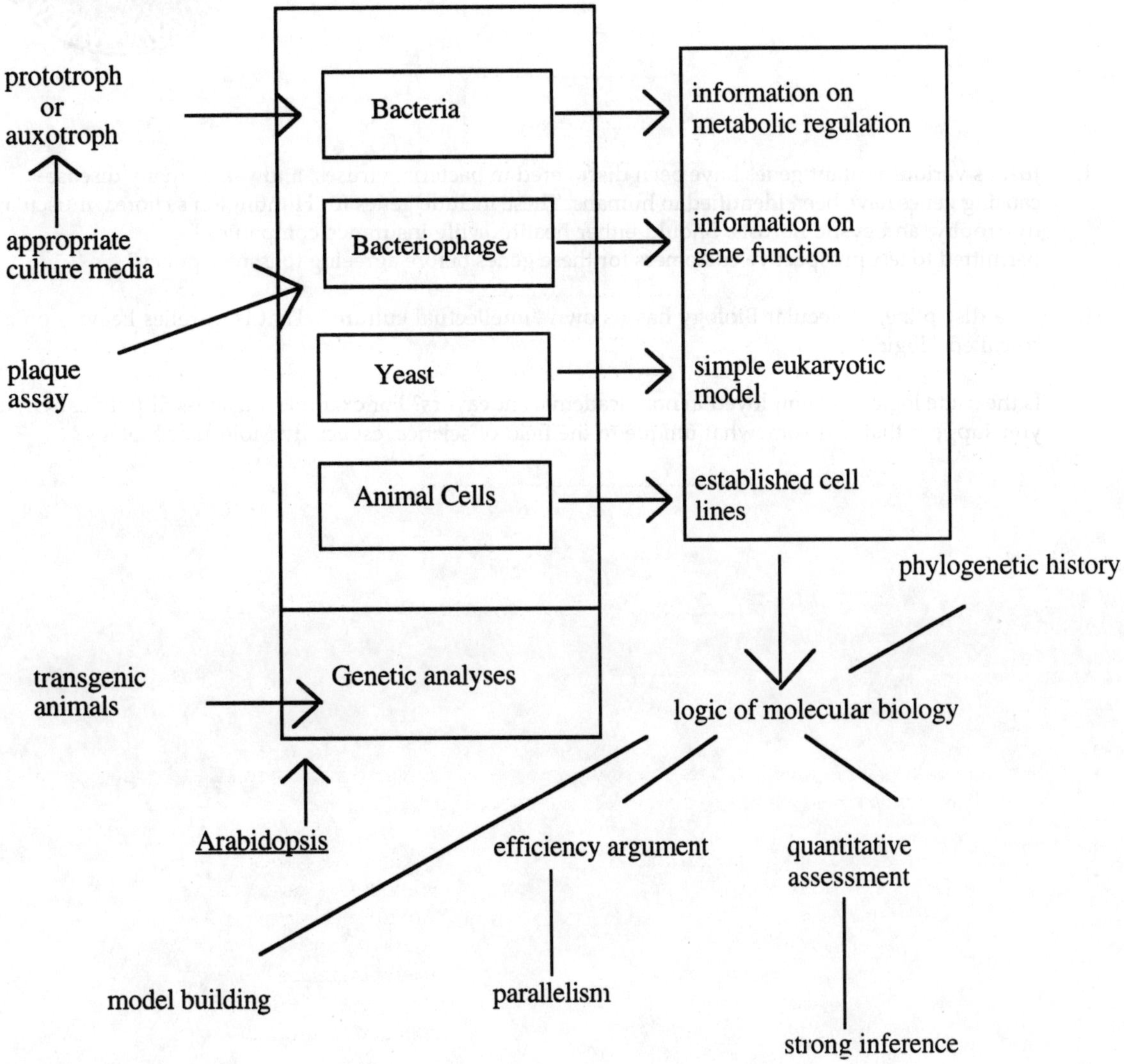
prototroph
or
auxotroph
appropriate
culture media
plaque
assay
Bacteria
Bacteriophage
Yeast
Animal Cells
Genetic analyses
information on
metabolic regulation
information on
gene function
simple eukaryotic
model
established cell
lines
phylogenetic history
transgenic
animals
logic of molecular biology
Arabidopsis
efficiency argument
quantitative
assessment
model building
parallelism
strong inference

Science and Society Issues

I. Just as various mutant genes have been discovered in bacteria, viruses, and yeast, many disease-causing genes have been identified in humans. These include genes for Huntington's chorea, muscular dystrophy, and cystic fibrosis. Should either health or life insurance companies be permitted to test prospective customers for these genes before agreeing to issue a policy?

II. As a discipline, Molecular Biology has its own "intellectual culture." That is, it relies heavily on a so-called "logic."

Is the same logic also employed in non-academic endeavors? For example, business or politics? Or, do you suppose that it is somewhat unique to the field of science, especially Molecular Biology?

2

Brief Review of Macromolecules

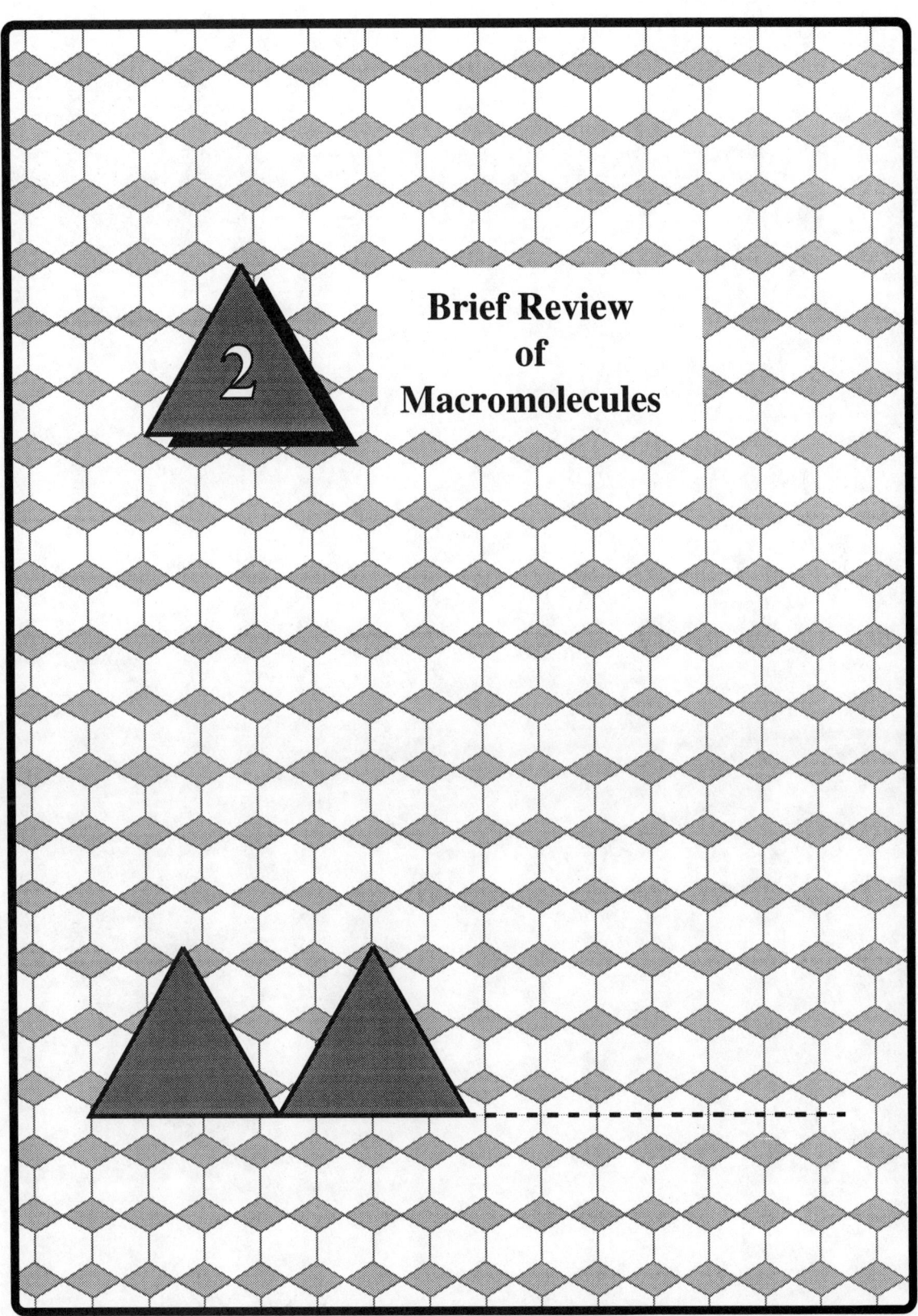

Here's Help

What gives amino acids their different characteristics?

There are 20 different amino acids common to biological systems. The only difference between the amino acids is the "side chain," with one exception, (proline). It is the side chain that gives each different amino acid its unique and characteristic properties. The amino acids can be classified into two general categories, hydrophobic and hydrophilic. Hydrophobic amino acids avoid direct contact with water molecules by clustering together, whereas hydrophilic amino acids interact freely with water molecules. Figure 2-2 of *EMB* illustrates the structures of the naturally occurring amino acids. Asterisks in that illustration denote the hydrophobic amino acids. Two amino acids, aspartic acid and glutamic acid, are acidic because they each contain a side chain with a carboxyl group. The carboxyl group can lose a hydrogen ion and, therefore, exist in the negatively charged form :

$$O{=}C{-}O^-$$

Conversely, three amino acids, lysine, arginine, and histidine, are basic. At physiological pH, a hydrogen ion is added to the amino group of the side chain to give the side chain a positive charge (NH_3^+). In the case of histidine, a hydrogen ion is added to the five-carbon ring. The charged amino acids are hydrophilic and reside on the outside of a completely folded protein.

How do amino acids give proteins their distinctive features?

This topic is discussed in detail in Chapter 4, so only a brief summary is provided here. The three-dimensional conformation of a protein largely determines its function. Enzymatic proteins are usually highly folded and compact. In contrast, structural proteins are usually elongated and sometimes fibrous in shape. For example, the spindle proteins in the mitotic apparatus of a dividing cell are structural proteins. Three-dimensional shape is determined by the way the various amino acids in the protein interact with water, shy away from water, or interact with each other. These various interactions are caused by hydrophilic amino acids, hydrophobic amino acids, and acidic-basic amino acids, respectively. Since each protein contains a unique amino acid sequence, each protein folds in a unique fashion. Hence, the large diversity in the function of proteins!

Do proteins and nucleic acids have common features?

In a general sense, *yes*. They are both biopolymers. That is, they are polymers that are naturally occurring in living cells. Both proteins and nucleic acids, since they are polymers, have a backbone structure. In proteins, the backbone consists of a chain of peptide bonds (see Figure 2-4 of *EMB*). In nucleic acids, the backbone consists of a chain of sugar-phosphate bonds (see Figure 2-7 of *EMB*).

Proteins and nucleic acids are similar in one additional sense. Side groups are covalently linked to the peptide or sugar-phosphate backbones. These side groups give the individual biopolymers their characteristic properties and functions.

Twenty different side groups are found in most proteins, whereas nucleic acids contain only four side groups [adenine, guanine, cytosine, and thymine (uridine in RNA)].

Let's review the sugar-backbone structure of nucleic acids!

Perhaps the easiest way to review this structure is to rotate Figure 2-7 of *EMB* 90 degrees. That is, examine the structure of a dinucleotide in a vertical orientation. In a slightly simplified form, it would look like this:

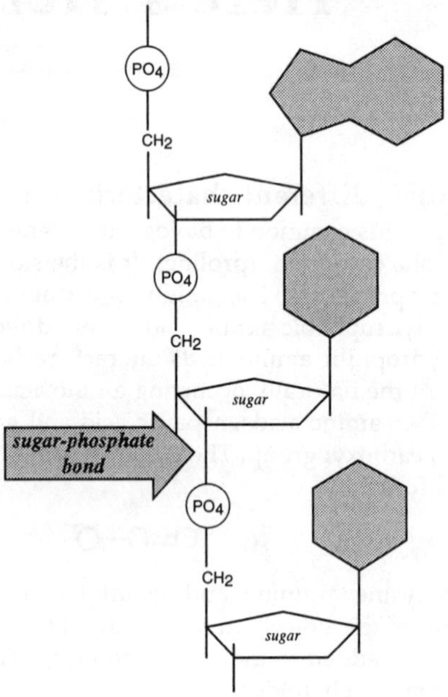

The purine and pyrimidine bases are sticking out from the sugars shown in Figure 2-7 of *EMB*. Please keep in mind that all nucleic acids, whether DNA or RNA, whether large or small, possess that same sugar-phosphate backbone. What distinguishes one nucleic acid from another is the length of the polymer and the distribution of the four bases along the backbone.

Shall we briefly review the weak forces that contribute so much to the shape of macromolecules?

Hydrogen bonds form when a hydrogen atom that is covalently bonded to an amino acid or nucleotide base is shared with another negatively charged atom (usually an oxygen, of a C=O or a nitrogen [N]). Hydrogen bonding between water molecules is extensive, as is hydrogen bonding between some amino acids and water molecules.

Hydrophobic bonds form between groups that, due to their inability to interact with surrounding water molecules, become clustered together. The strength of the hydrogen bonds that form between water molecules is sufficiently strong and pervasive to drive closely spaced hydrophobic groups together. A water shell forms around the hydrophobic groups and stabilizes them.

Ionic bonds occur between oppositely charged groups. These bonds are by far the strongest of the noncovalent bonds. They are often responsible for the binding of one protein to another, or the binding of proteins to nucleic acids.

Van der Waals attractions represent the weakest of the noncovalent bonds and are important only when two atoms are very close together. When two molecules lie close together, transient fluctuations in electron distribution are often induced. This exchange of electrons causes momentary charge asymmetries. For that instant in time, the atoms move closer together, forming a weak cohesive force.

Van der Waals attractions represent the weakest of the noncovalent bonds and are important only when two atoms are very close together. When two atoms lie close together, transient fluctuations in electron distribution are often induced. This shifting of electrons causes momentary charge asymmetries. For that instant in time, the atoms move closer together, forming a weak cohesive force.

Why are weak bonds and forces so prevalent among macromolecules?

First, let's comment on the strong bonds—that is, the covalent bonds that constitute the backbone of biopolymers such as proteins and nucleic acids. The covalent bonds in backbones are 10-20 times stronger than the weak bonds just described. They are so strong that a backbone structure seldom breaks; despite the fact that proteins and nucleic acids are occasionally moved around in the living cell, or, in the case of some proteins, even stretched or compressed.

Now let's consider weak bonds. Most biopolymers, in order to carry out their function as effectively and efficiently as possible, contain bends, twists, turns, folds, or even convolutions in their 3-D structure. This enables them to tightly interact in a "lock and key" fashion. To enhance effectiveness, biopolymers sometimes undergo subtle shape changes during side chain interactions. The weak bonds are easily broken and reformed at the temperatures and salt concentrations found in living cells. This "doing" and "undoing" permits shape changes to occur easily. Covalent bonds, on the other hand, are usually stable under normal physiological conditions and require specific enzymes in order to be broken.

For example, during replication, the two strands of DNA unwind, are replicated by various enzymes, and then wind back together. The hydrogen bonding between the two strands is relatively easy to undo. If the two strands were held together by covalent bonds, each time single strands of DNA were required for a particular process (e.g., replication or RNA synthesis), an input of a tremendous amount of energy, as well as the on-site presence of highly specific enzymes, would be required.

We will return to these weak bonds in detail when we consider nucleic acids in Chapter 3 and proteins in Chapter 4.

What is gel electrophoresis?

Electrophoresis is a technique that makes use of the different charges, sizes, and shapes of molecules in order to separate them in an electric field. (Note that by size we mean molecular weight). Our discussion here is limited to electrophoresis done in polyacrylamide gels using DNA and protein samples.

An electric field is run across a polyacrylamide gel and molecules are allowed to migrate through the gel. Naturally, molecules with positive charges move toward the negative pole and molecules with negative charges will move toward the positive pole. However, the gel physically limits the movements of molecules. Therefore, smaller molecules will tend to move faster and farther than larger molecules of the same charge. This is the basic concept behind separation by electrophoresis.

Having stated that basic principle, let's look at the separation of DNA molecules by electrophoresis. Remember, all DNA molecules are extremely negatively charged since the backbones are made up of negative phosphate groups. Also, all DNA molecules have the same shape: the double helix. Therefore, electrophoresis is employed to separate DNA molecules based on size difference (again, by size we mean molecular weight). This is done by putting the sample at the negative end of the gel. All molecules in the sample will, of course, migrate away from the negative end and toward the positive end. Smaller DNA molecules will migrate farther than will larger, more cumbersome DNA molecules, as is shown in Figure 2-13 of *EMB*.

The situation is a little more complicated for proteins. With DNA, the only difference between molecules is size. That makes separation easy. However, proteins vary not only in size, but also in shape and charge. All proteins do not even migrate in the same direction, as some have a negative charge, some have a positive charge, and some are neutral. To accomplish a successful separation, we need to equilibrate as many factors as possible. First, the proteins are denatured, usually by using heat or 8M urea.

β-mercaptoethanol (βME) is then used to break disulfide bonds. This unfolds the proteins further so that they are all approximately equal in shape. Electrophoresis can then be used to separate the molecules based mostly on charge (actually, on the ratio of charge to mass).

In order to enhance the separation of proteins in a *complex* mixture, a more sophisticated procedure that first separates proteins on the basis of charge is directly coupled with a method that separates on the basis of size. This is called two-dimensional gel electrophoresis. As the name indicates, two levels of separation are combined to give more sophisticated results. Figure 2-11 of *EMB* illustrates this method. The first dimension gel is rotated 90 degrees and laid directly on top of the second dimension gel. The second dimension gel contains a detergent (SDS) which dissociates larger proteins into their subunits. Thus, the first dimension separates mainly on charge differences, while the second dimension separates mainly on size differences.

Key Terms

Peptide bond

Disulfide bond

Ionic and covalent bonds

Hydrogen bonds

Phosphodiester bond

Van der Waals force

Random coil

Nucleoside

Nucleotide

Amino and carboxyl termini

Base stacking

Hydrophobic

Hydrophilic

Electrophoresis

For More Information . . .

Where can I easily get other explanations on these topics?

A discussion and diagrams of amino acids and nucleotides can be found on pp. 56-59 of *Molecular Biology of the Cell*, 3rd edition, by Bruce Alberts, *et. al.* A good explanation of electrophoresis can be found on pp. 169-170 of the same book.

Both covalent and noncovalent bonding is described in detail on pp. 129-141 in Watson *et. al.*, *Molecular Biology of the Gene*, 4th edition.

The Wise Owl Says

If you are willing to learn chemical formulas (e.g., Figure 2-6 of *EMB*) and work on understanding the nature of basic chemical reactions, like weak interactions that give macromolecules their shapes, you will be better equipped to perform research in the areas of high technology of molecular biology (e.g., protein structure). That is because you will truly understand the chemical basis of the methods (e.g., electrophoresis) which are routinely employed.

Exercise

I. By examining Figures 2-3 and 2-7 in *EMB*, you will realize that proteins and nucleic acids, as well as virtually all other biopolymers, are synthesized by a repetitive dehydration reaction. In other words, the covalent peptide or phosphodiester bond is formed by the elimination of a water molecule. The actual mechanisms, or enzymes, involved in dehydration reactions will be considered in detail in later chapters of *EMB* (proteins = Chapter 9; nucleic acids = Chapters 7 and 8).

Please draw the polymerization reaction for protein synthesis, illustrating the dehydration.

Likewise, draw the nucleotide polymerization for DNA synthesis, illustrating the dehydration involved in phosphodiester bond formation.

A fundamental difference exists between the two polymerizations. The energy for the formation of the phosphodiester bond in DNA is provided by the "incoming" nucleotide: an energy rich nucleoside triphosphate. In contrast, the energy for the formation of the peptide bond in protein comes largely from the hydrolysis of a nucleoside triphosphate that is brought in, along with the incoming amino acid, to the elongating polypeptide. It does not, however, become a permanent part of the protein.

However, the breakdown, by hydrolysis, of proteins and nucleic acids is similar for both types of molecules. In both cases, a water molecule is added to either the peptide or phosphodiester bond.

Please draw the hydrolysis of a dipeptide and a dinucleotide. Illustrate the addition of a water molecule to the relevant components of the preexisting peptide or phosphodiester covalent bond.

II. Draw the chemical structures of the side groups for the following segment of a protein molecule. Illustrate the hydrogen bonding, ionic bonding, and hydrophobic interaction that occurs between the appropriate amino acids.

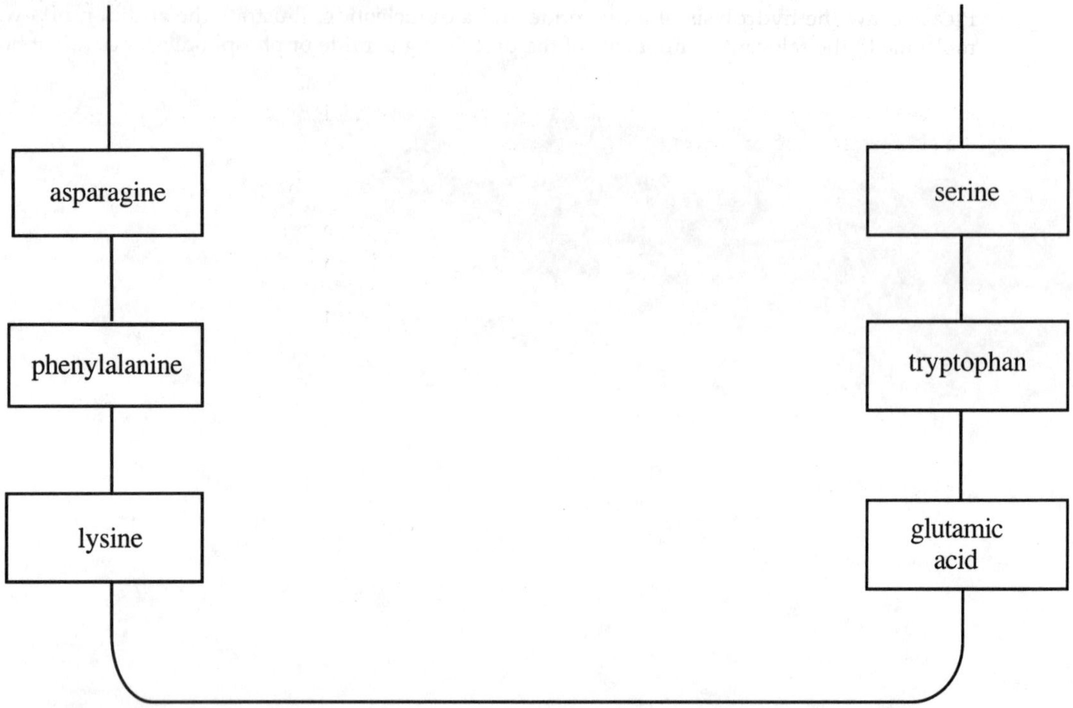

Tough Nuts

I. Please draw out the chemical formulas for the nine amino acids that are found in proteins but cannot be synthesized in human cells. They include:

valine	lysine
leucine	phenylalanine
isoleucine	tryptophan
threonine	histidine
methionine	

Please speculate as to why these amino acids cannot be synthesized from normal components of the human diet.

This is indeed a tough nut. Does a comparison of the formulas give any clues? Some amino acids, like cysteine and proline, found only rarely in proteins are synthesized by human cells. Others, like valine and leucine, are relatively common in proteins and yet cannot be synthesized by human cells. Does this knowledge help crack this tough nut? Where does the concept of evolutionary change fit into your thinking? How did it come to pass that those 20 amino acids, and only those 20, are found in proteins? An organic chemist could easily synthesize dozens of additional simple amino acids, similar to the 20 naturally occurring ones illustrated in Figure 2-2 of *EMB* . Perhaps the same forces that led to the choice of those 20 operated to establish which amino acids human cells can synthesize and which they can not.

II. Due to the strength of the covalent bonds that comprise a biopolymer's backbone, macromolecules are remarkably stable under normal conditions in a living cell. Biopolymers can last indefinitely and occasionally outlive the cells that contain them.

Yet it is often in the best interest of a cell to change its inventory of polymers. Please speculate on ways the living cell might go about eliminating surplus enzyme molecules or excess nucleic acids (e.g., ribosomal RNA).

You will probably be impressed by the complexity of this issue. Could it be that the mechanisms employed by cells to change particular polymers are even more complex than the mechanisms employed to synthesize the polymers in the first place?

No doubt some of the mechanisms you suggest will involve the breaking of covalent bonds. Where will the energy come from for bond breakage?

Study Questions

1. What is the difference between a nucleoside and nucleotide?

2. Which nucleic acid base is unique to DNA? to RNA?

3. Draw a polypeptide that exhibits three different types of noncovalent interactions.

4. Draw a 4-base pair segment of a DNA molecule, including each nucleotide and the associated bonds involved in the maintenance of the double helix.

5. What is the difference between covalent, ionic, and hydrogen bonds?

6. Discuss the difference between DNA and RNA and the effect this difference has on stability. Suggest how this fact relates to the functions of the two macromolecules in an organism. (Hint: the "difference" is not that one is double-stranded and the other is single-stranded.)

7. Why can't DNA be the template that *directly* orders amino acids during protein synthesis?

8. Suppose a mixture of proteins yields 15 bands on the first dimension of a 2-D electrophoresis system, as shown in Figure 2-11 of *EMB*. Why can one virtually always expect the second dimension to reveal more proteins (e.g., approximately 60 in Figure 2-11) than the 15 observed in the first dimension?

Answers to Chapter 2 Study Questions

1. A nucleotide is a nucleoside (base + sugar) with a phosphate group linked to the 5' carbon of the sugar.

2. DNA, thymine; RNA, uracil

3.

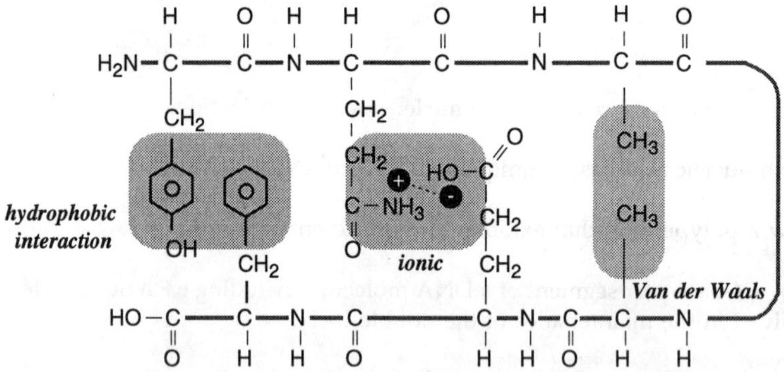

4. See various illustrations in the textbook (*EMB*).

5. Covalent bonds are the only ones that involve electron sharing. Ionic bonds involve attraction between unlike charges. Hydrogen bonds are formed between either the oxygen atom of a carboxyl group or the nitrogen atom of an imine group and a hydrogen atom of either an amine group or a carboxyl group.

6. The sugar group of DNA is deoxyribose and that of RNA, ribose. The two sugars differ only in the group attached to the 2' carbon—a hydrogen in deoxyribose and a hydroxyl (OH) group in ribose. The presence of the hydroxyl group makes ribose more prone to degradation, thus RNA is short-lived relative to DNA. This situation is well suited to the function of RNA which serves predominantly as a messenger that must be broken down quickly (in many cases). DNA's relatively high stability is ideal for preserving the genetic code.

7. In eukaryotes, the DNA is separated from the site of protein synthesis (cytosol) by a nuclear envelope that prohibits the passage of large molecules such as proteins.

8. Proteins differ substantially in size as well as charge. Yet the first dimension (pH gradient) separated exclusively according to charge. The sieving of the second dimension would reveal more proteins, because size alone (due to SDS) is the basis of this separation.

Concept Map

Shall we "network" this chapter together into a concept map?

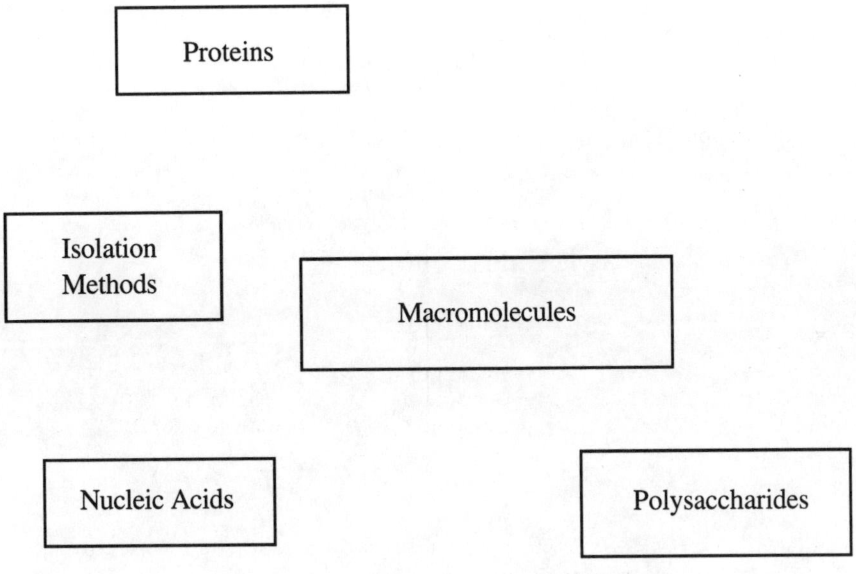

Please add arrows or connecting lines to illustrate the relationship between the above boxes and the following terms:

amino acids
nucleoside
polypeptide chain
phosphodiester bond
hydrogen bonding
ionic bonds
random coil
hydrophobic interaction
DNA cloning
Van der Waals attraction
gel electrophoresis

3 Nucleic Acids

Here's Help

What holds the double-stranded DNA molecule together?

In a molecule of double-stranded DNA, the base pairs are separated by 3.4 angstroms (0.34 nanometers). Since one turn of the helix is comprised of 10 base pairs, one turn measures 34 Å (3.4 nm) in vertical displacement. The helix has two grooves, called the major and minor grooves. The sugar-phosphate backbones form the "outsides" of the grooves, while the base pairs are held on the inside. This is illustrated in Figures 3-1 and 3-6 of *EMB*. *Please note*: Some textbooks use "angstrom" units, while others use nanometers for DNA/protein measurements.

One of the beautiful features of DNA as the genetic material is the perfect complementarity of the two strands. If at a given point on one strand there is an A, then at the same point on the other strand there always is a T. If an error in base sequence should occur in one strand, then the cell should be able to restore the correct sequence as long as the other strand remains intact. This is like a library having an extra copy of every one of its books!

The "double-strandedness" of a DNA molecule is maintained by hydrophobic interactions and hydrogen bonding. The hydrophobic bases "prefer" to be shielded from water. In order to achieve this, the bases of two complementary strands come together, forming the familiar helix, so that the negatively-charged phosphate groups of the backbone are on the outside and the bases are on the inside.

Two hydrogen bonds form between an A and a T, three form between a G and a C. Thus, DNA breathing, a phenomenon in which double-stranded regions open to become single-stranded "bubbles," occurs more often in A-T rich regions since the two hydrogen bonds in A-T base pairs break more easily than the three hydrogen bonds found in G-C base pairs.

While hydrogen bonding and hydrophobic interactions stabilize the DNA molecule, "Coulombic repulsion" among the phosphate groups of the backbones, tends to destabilize it. These phosphate groups are very electronegative, and the repulsive forces generated by their close proximity in the double-stranded helical form are strong enough to cause the strands to separate. However, under normal conditions in a living cell, these charges are neutralized by cations.

What is *denaturation*, anyway?

When the forces holding the two strands of DNA together are overcome, the strands separate and the molecule is said to be **denatured**. One way to denature DNA is to lower the salt concentration of the solution. Without a sufficient concentration of cations to neutralize the phosphate groups, the molecule separates into its two component strands. Another more practical way to denature DNA is to heat it. As the temperature increases, so does the energy of thermal motion. When the energy of thermal motion becomes greater than the cumulative energy of the hydrogen bonds between the base pairs, the two strands separate. Obviously, the more hydrogen bonds, the higher the temperature required to denature the molecule. For different molecules that contain a similar number of base-paired nucleotides, the one with the highest G-C content will denature at the highest temperature. This is because G-C base pairs contain three hydrogen bonds, while A-T base pairs contain only two.

The more "open," or denatured, a DNA molecule is, the more ultraviolet light it can absorb. This basic fact can be employed to figure out the relative amount of A-T base pairs, or G-C base pairs, contained in any given DNA molecule. This can be done by analyzing the "melting curve" data procured from a solution of DNA. A melting curve is obtained by plotting the A_{260} (the amount of light absorbed with a wavelength of

260 nm) against the temperature. The point on this curve at which the increase in absorbance is half complete (when 50% of the DNA in the solution is denatured) is called the T_m. Each different DNA molecule has a unique and characteristic T_m!

While denaturation can yield useful information about a DNA molecule, so can renaturation (the reconstitution of the double-stranded condition). A large DNA molecule that has been fragmented and denatured can be allowed to slowly renature. Please recall that smaller molecules will renature faster than larger molecules when all other factors, especially concentration (g/ml), are held constant. This is because the molar concentration of the smaller molecules will be higher and will therefore result in more collisions.

The fraction which remains denatured at any point in time can be plotted on graphs in various ways. These denaturation-renaturation analyses provide information about the DNA molecule such as the number of different repeated sequences present in the molecule, the number of copies of each sequence, and the number of base pairs contained in each of those sequences.

Want a stepwise lesson in DNA structure?

The following illustrations progress from the very simple to the more complex. Please study them, one at a time, and develop your understanding of DNA structure:

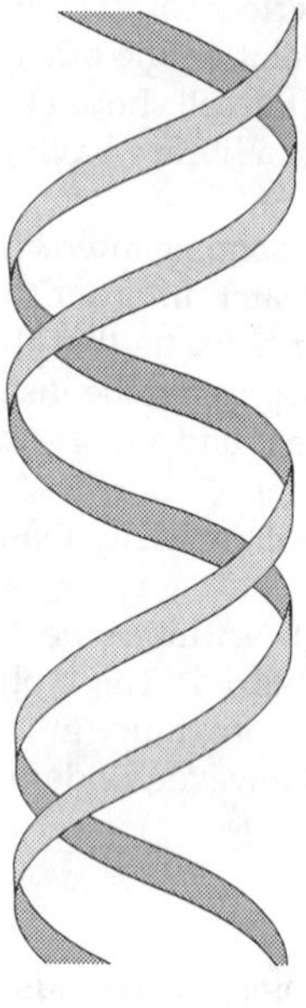

This diagram illustrates the double-stranded feature of DNA

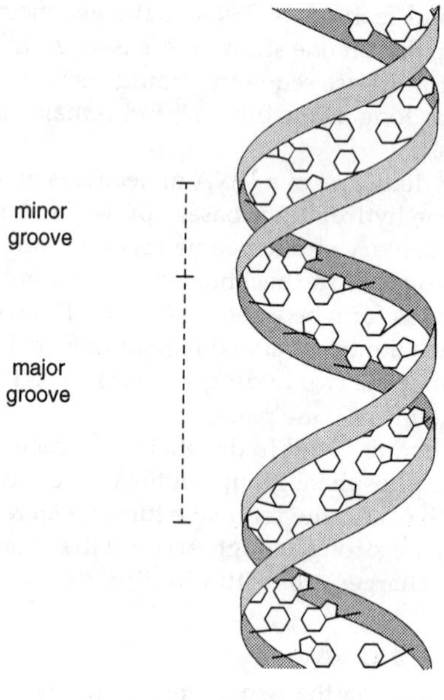

Please use this diagram (which is similar to the one in Figure 3-1 in *EMB*) to visualize the location of the bases, inside the helix where they are protected from environmentally induced damage, and the minor and major grooves

This diagram highlights the sugar-phosphate backbone, from which the bases (adenine, thymine, guanine, cytosine) protrude.

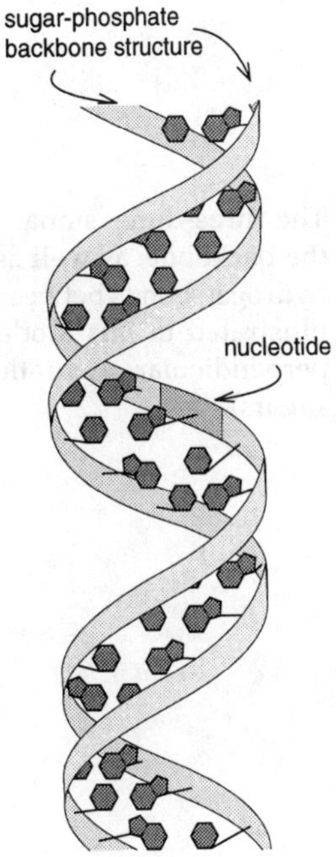

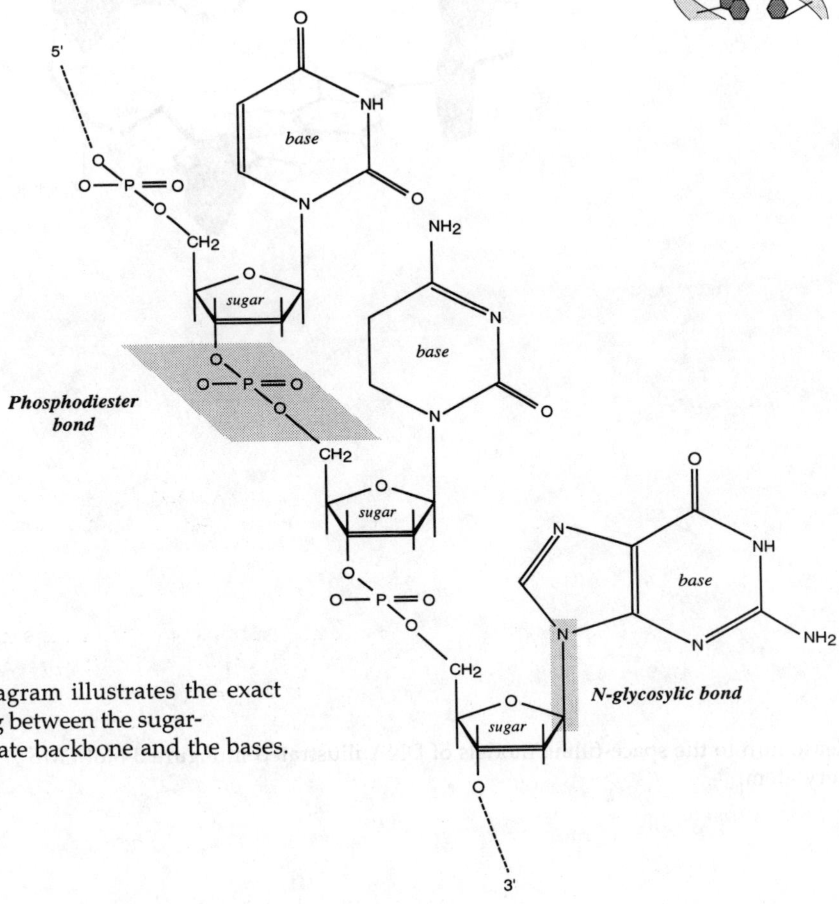

This diagram illustrates the exact bonding between the sugar-phosphate backbone and the bases.

The three-dimensional features of the backbone, as well as the hydrogen bonds between bases, are illustrated in this model. (Note the perpendicular orientation of the sugars).

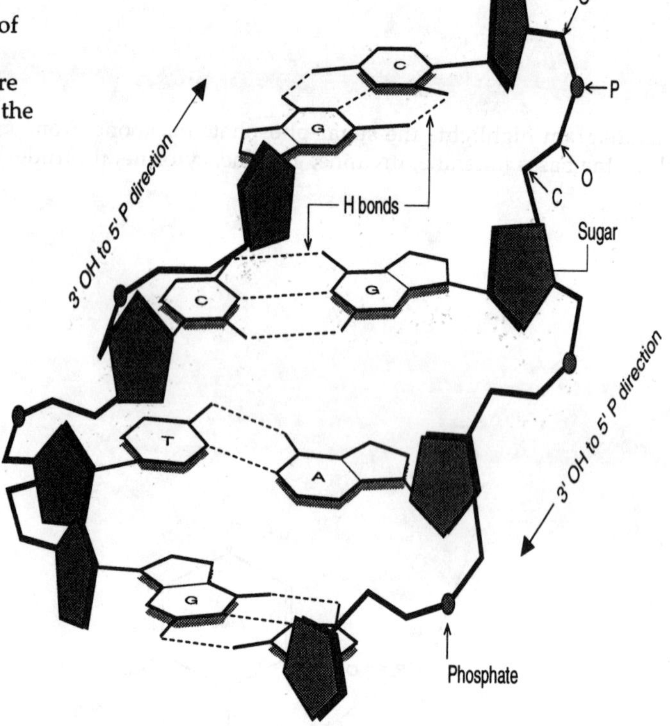

Now, please turn to the space-filling models of DNA illustrated in Figure 3-6 of *EMB* . They illustrate each and every atom!

Here's more information on A, B, and Z-DNA!

The DNA molecule just constructed in a stepwise fashion represents the now famous, conventional model deduced by James Watson and Francis Crick from Rosalind Franklin's x-ray diffraction data over 35 years ago. This conventional model is thought to represent the "native" structure of DNA. This native DNA structure, as it exists in most living cells, is termed the "B-conformation."

Recently, several alternative conformations of DNA have been discovered. Two of them, the "A-conformation" and "Z-conformation," may actually exist in living cells. Indeed, the B-form may convert to either the A- or the Z-form to adapt to the current situation within a particular cell.

Here is how the three forms differ:

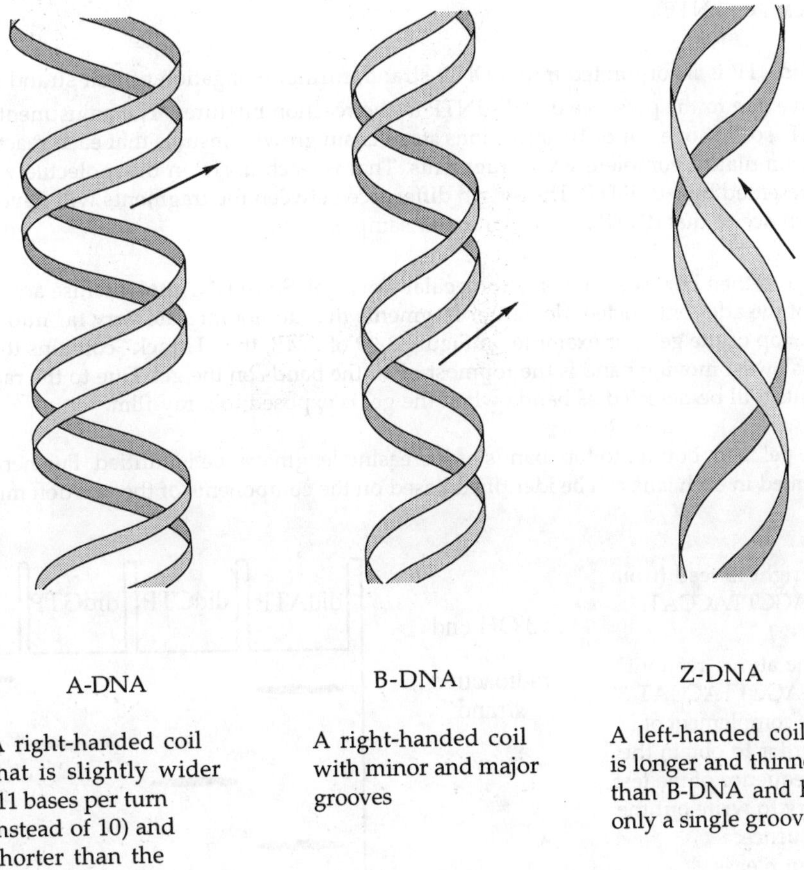

A-DNA

A right-handed coil that is slightly wider (11 bases per turn instead of 10) and shorter than the native, B-DNA

B-DNA

A right-handed coil with minor and major grooves

Z-DNA

A left-handed coil that is longer and thinner than B-DNA and has only a single groove

What might cause the transition of the natural B-form to either the A- or Z-form? It is believed that the A-form is a somewhat dehydrated version of the B-form. The Z-form, however, is thought to be an enzymatically modified version. The major shift in conformation is thought to be caused by the enzymatic addition of methyl groups to cytosine and guanine.

Here's a summary of the "dideoxy" sequencing method which is described in more detail in Figure 3-18 in *EMB*:

Many copies of the single-stranded DNA to be sequenced are annealed to a short DNA primer, (approximately 20 base pairs long) and divided into 4 portions. To each portion the enzyme DNA polymerase, a complete set of the four deoxynucleotide triphosphates (dATP, dTTP, dCTP, and dGTP), and a radioactive tracer (e.g., ^{32}P-dATP) are added. Then, a tiny amount of one of the four dideoxynucleotide triphosphates (didNTP) is added to each of the four tubes. That is, the first tube contains a complete reaction mixture plus didATP; the second tube contains a complete reaction mixture plus didTTP, and so forth.

A key point: The ratio of dNTP (i.e., dATP, dTTP, dCTP, and dGTP) to didNTP is predetermined so that a didNTP is incorporated into the newly synthesized strand of DNA approximately once in every 100 regular nucleotides (i.e., dNTP).

Whenever a didNTP is incorporated into a DNA strand, further elongation of that strand stops! Each strand is radioactive due to the presence of ^{32}P-dNTP in the reaction mixtures. The adjustment of the ratio of didNTP to dNTP, so that one out of 100 insertions stops chain growth, insures that each reaction mixture will eventually accumulate a complete set of fragments. That is, each dNTP in the molecule will eventually be represented by a didNTP. The length difference between the fragments will depend upon the frequency of occurrence of that dNTP in that particular sample.

The fragments can then be visualized on a molecular sieve gel. Since DNA polymerase adds nucleotides to the free 3' OH of the adjacent nucleotide, larger fragments that do not migrate very far into the gel reside closer to the top of the gel. For example, in Figure 3-19 of *EMB*, the "T track" contains the largest fragment since its slowest moving band is the topmost of all the bands on the gel. Due to the radioactive labels, the fragments will be detected as bands when the gel is exposed to x-ray film.

By reading the gel from bottom to top, bands of increasing length can be identified. Furthermore, the nucleotide represented in each lane can be identified based on the components of the reaction mixture from whence it came.

The gel to the right is read (from the bottom up) as AGCTTACGAT.

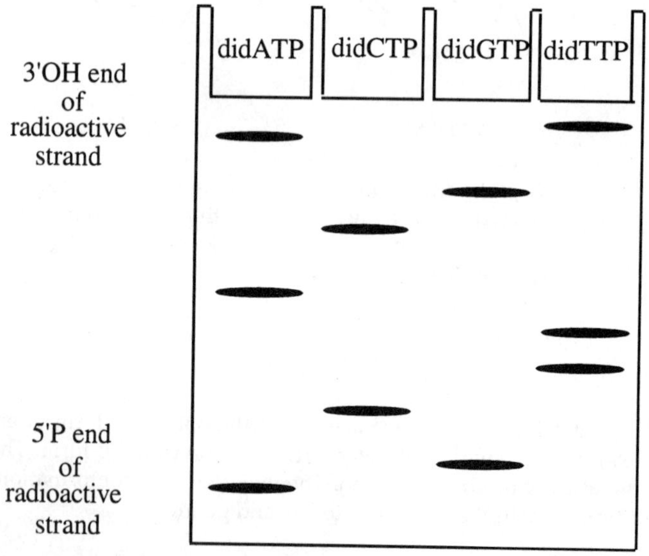

A key point: The above readout represents the 5'P-AGCTTACGAT-3' OH sequence of the complement of the test DNA! In order to obtain the actual nucleotide sequence of the test DNA, it is necessary to write out the complementary sequence: TCGAATGCTA. But please remember, the DNA strands are antiparallel, so the above sequence (TCGAATGCTA) represents the 3'OH to 5'P sequence. By convention, nucleotide sequences are written in the 5'P to 3'OH direction. So, it is necessary to "flip" (i.e., invert) the above sequence: 5'P-ATCGTAAGCT-3'OH. This last sequence finally represents the actual nucleotide sequence of the test DNA!

Key Terms

Major groove

Minor groove

Base composition

Satellite DNA

B-DNA

A-DNA

Z-DNA

Supercoiling

Cruciform

Palindromic

Topoisomerases

Native

Denaturation

Renaturation

A_{260}

T_m

Hyperchromic

Hypochromic

Copy number

Filter hybridization

Heteroduplex mapping

Endonuclease

Exonuclease

DNA gyrase

Dideoxynucleotide

For More Information . . .

Where can I easily find some other explanations on these topics?

DNA structure: Try Alberts *et. al., Molecular Biology of the Cell,* 3rd edition: pp. 100-101. Also try Watson *et. al., Molecular Biology of the Gene,* Figure 9-2 (p. 242) and Figure 9-9 (p. 250).

DNA Sequencing: Having trouble visualizing the Sanger method? Look at Figure 7-7 in Alberts *et. al.* (3rd edition), or try Figure 9-45 (p. 276) in Watson *et. al.*

Hybridization: An excellent diagram is shown in Figure 4-13 of Alberts *et. al.* (3rd edition). Also try Figure 19-15 (p. 609) in Watson *et. al.*

The Wise Owl Says

"Science is a Human Endeavor"

Science is practiced by people who have the same emotions, drives, likes, and dislikes as you. Although the subject matter, for example, a test tube of purified DNA, is itself often detached from human feelings, the scientist is not.

When the stakes are thought to be high, for example when a Nobel prize might be awarded for a specific discovery, the pace of research quickens, competition increases, and personality conflicts that lurk below the surface sometimes emerge.

Want To Read About Such Competition?

James D. Watson shared the Nobel prize with Francis Crick for elucidating the molecular structure of DNA depicted in Figures 3-1 through 3-4 of *EMB*. He described the interpersonal relations he experienced during the course of making his important discovery in his book: *The Double Helix; a Personal Account of the Discovery of DNA* (Watson, J.D., Atheneum press, N.Y., 1968). Fascinating reading!

Science magazine [(8 April 1988) 240 :141-144] contains a fascinating account of the isolation of the cystic fibrosis gene. Amazing race!

Exercise

I. Antiparallel orientation of complementary DNA strands

Let's expand the illustration of the antiparallel orientation of complementary strands of a double helix, shown in Figure 3-4 of *EMB*. Draw the complete chemical structures of 2 sugar-phosphates on each side of the double helix.

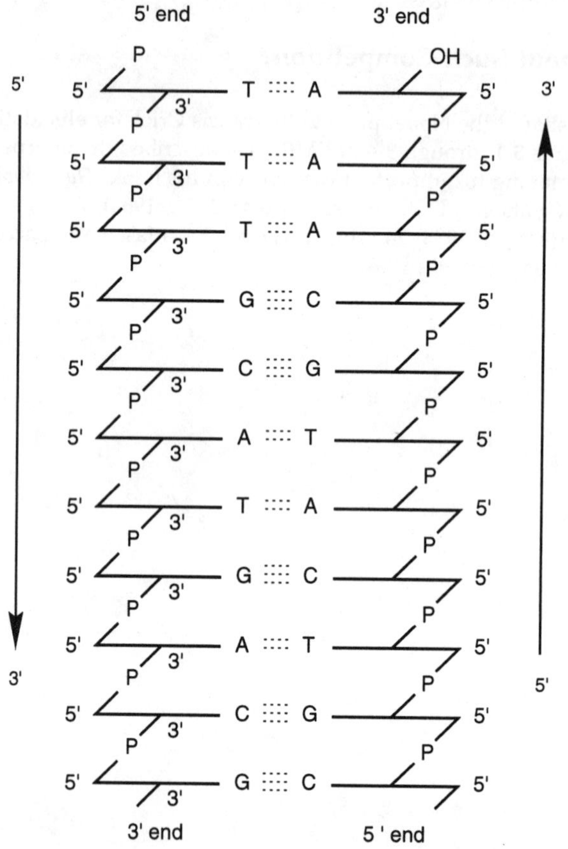

draw 2 sugar phosphates

draw 2 sugar phosphates

P.S.: The last page of this chapter shows a completed illustration.

II. Individual DNA Models

Let's prepare a DNA model that represents your own nucleotide sequence! We will draw a nucleotide sequence, based on your social security number (repeated twice), on a paper bag. The actual sequence you construct will be 10 base pairs long.

1. Convert your social security number into a nucleotide sequence as follows:

A=0,1,2	Example:	026-31-5894 converts (twice) into:
T=3,4,5		AAGT/ATCCTAAGTA/TCCT
G=6,7		You will draw the middle ten bases
C=8,9		

2. Cut a coil out of a paper bag to make the following shape (sizes are approximate):

3. Write in the *complete* chemical structure (sugars, phosphates, and bases) of the middle ten nucleotides. Those ten nucleotides should be drawn proportionately, representing one full turn (34 A) of the double helix. Refer to Figures 2-6, 2-7. and 3-2 in *EMB*; ALL atoms (C; O; N; P; H) should be present and accounted for.

4. Please be certain the following features can be easily recognized:
   a. the antiparallel nature of dsDNA
   b. H bonding
   c. the sugar-phosphate backbone
   d. exposed phosphates with negative charges

5. Calculate the T_m of your DNA, using the graph shown below:

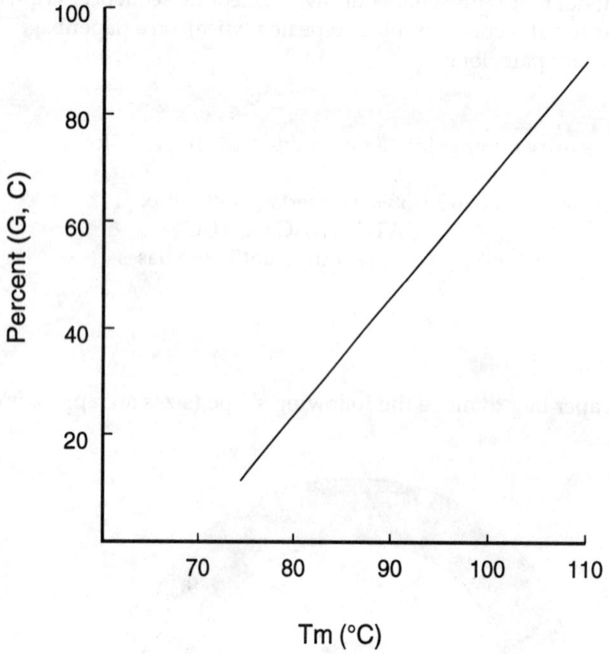

III. Supercoiling of DNA

Double-stranded DNA is usually either positively or negatively twisted upon itself. Both forms are found in living cells. This phenomenon is called supercoiling, and perhaps the best way to understand it is to twist a rubber band!

Here's how:

Cut a rubber band.* Hold one end firmly between your fingers and twist the other end one full turn clockwise, then hold the two ends together. Now notice how the circular band naturally twists upon itself to form a figure eight. By forming a supercoil the strain introduced on the clockwise turn is relieved. Conventionally, this twist is called a positive, or clockwise coil. The diagrams on the next page illustrate this exercise.

*A short, wide rubber band works better than a long, narrow one.

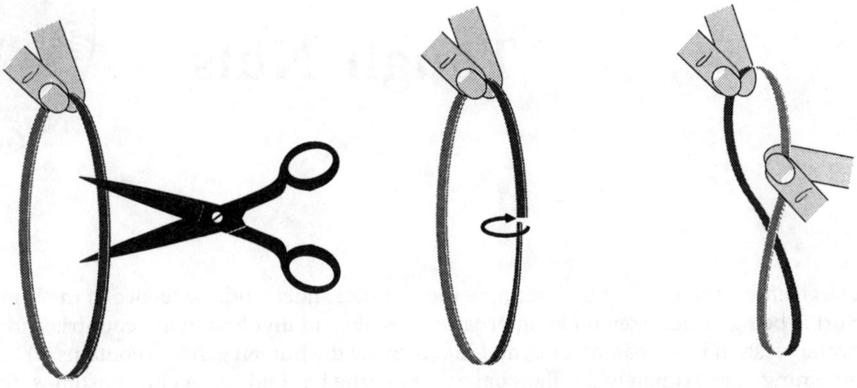

In a living cell, both supercoiling and its reverse, relaxation, are generated by the actions of specific enzymes called topoisomerases. Type II topoisomerase, more commonly called gyrase, is just one example of a topoisomerase that specifically introduces negative supercoils into DNA. As you might imagine, all this twisting of large DNA molecules requires a lot of energy. This energy is mainly supplied by the hydrolysis of ATP.

Supercoiling is especially important during DNA replication (Chapter 7 of *EMB*). Unwinding the DNA at the replication fork introduces positive supercoiling that needs to be relaxed by enzyme-mediated negative supercoiling. This job is usually done by gyrase; otherwise, a kink would develop at the replication fork, and nucleotide polymerization would be inhibited.

Tough Nuts

I. A large effort is underway to determine the complete nucleotide sequence of the human genome. This effort is being undertaken on an international scale and involves many countries, including the USA, several western European nations, and Japan. Since the human genome contains 100,000 genes or more, containing approximately 3 billion base pairs in the haploid set of chromosomes, the project is a formidable one. The goals of the project are to uncover as much information about human genes as possible and to map the exact locations of each gene. The project will be somewhat complicated due to variation in the nucleotide sequence among different individuals.

To what extent will a complete nucleotide sequence of the human genome be interpretable and useful?

Please consider the following points:

a. Will the causes of various disease syndromes (e.g., Alzheimer's disease and polycystic kidney disease) be understood from a complete nucleotide sequence of the relevant gene?

b. Much of the human genome has no apparent function. That is, it does not code for proteins and is often referred to as "junk DNA," or "filler DNA." Why bother to sequence this DNA since sequencing is a very expensive procedure?

c. The future of molecular medicine, especially those aspects that deal with attempts to design and synthesize "custom tailored" drugs to fight disease, requires as much nucleotide sequence information as possible. Pharmaceutical companies will deduce protein structures from nucleotide sequences and synthesize modified versions that will be able to replace the defective proteins in diseased patients.

II. **How many different configurations (e.g., A, B, Z) for DNA are there?**

It is well-known that DNA molecules can exist in more than one double helical configuration. In living cells, DNA is believed to exist largely in the B-configuration. Some of the properties of the various forms, including the number of base pairs per turn of the double-stranded helix (9-12), are strikingly different. These physiological differences are potentially very important to molecular research. Researchers want to know if DNA in the living cell changes conformation. If indeed it does, do these changes alter the function of that DNA? Is it possible that some DNA–protein interactions occur preferentially with one specific DNA conformation? Could these special interactions and conformations generate specific gene expression patterns (e.g., during development of the embryo)? Could they serve to stabilize, or protect, DNA in cells that are largely metabolically dormant (e.g., bone cells)?

At least two considerations come into play here. First, the B-form predominates, so it will be difficult to isolate the trace amounts that exist in the alternate conformation. Extreme caution must be taken to insure that the isolation conditions don't cause a switch from the rare form to the favored B-form.

Second, it will be difficult to predict the function in the living cell of conformations that are generated in cells used during laboratory experimentation. Recall the warning of the wise owl about the limitations of molecular biology and its lack of predictive powers!

Propose a way to crack these tough nuts!

Far Out

I. In this chapter an additional feature, FAR OUT, has been introduced. It will appear from time to time and is intended to present possible avenues for future research, avenues that are just now being explored. Some will develop well; others will fail miserably. As a discipline, molecular biology has enjoyed spectacular achievement, with the successes far outweighing the failures. Optimism is, therefore, warranted. Although there are always numerous reasons for being pessimistic, the pessimists have usually been proven wrong in molecular biology.

In fact, it is often said that congenital pessimists should not pursue careers in science. Pessimists exist in all walks of life. Here is a list of some famous ones that were proven wrong!

Lord Kelvin, President, Royal Society, London (1895): "Heavier than air flying machines are impossible."

Charles H. Duell, Director of U.S. Patent Office (1899): "Everything that can be invented has been invented."

Tris Speaker, baseball player (1921): "Babe Ruth made a big mistake when he gave up pitching."

Robert Millikan, Nobel Laureate in Physics (1923): "There is no likelihood man can ever tap the power of the atom."

FAR OUT will provide brief sketches of research projects that will certainly not appeal to pessimists. However, these projects may become more realistic by the time *EMB* students begin their careers in science.

II. In the past few years, several laboratories have succeeded in synthesizing oligonucleotide polymers that anneal to *double-stranded* DNA, to form "triplex DNA."* These synthetic DNAs, that hybridize best to normal double-stranded DNA, are usually rich in T and C (pyrimidine bases) and are approximately 20 nucleotides long.

It is predicted that the "triplexing" of DNA will eventually be useful both as an experimental tool and as a therapeutic agent. In the former instance, it is imagined that synthetic oligonucleotides that contain covalently attached cutting chemicals (e.g., oxidizing agents) could be used to immediately cut double-stranded DNA at specific nucleotide sequences upon direct contact with those sequences. In the latter case, it has been suggested that synthetic oligonucleotides that are custom designed to hybridize with sites on DNA where transcription is initiated could be used to shut down viral gene expression. Perhaps by using this method effective anti-AIDS virus drugs could be produced.

*See *Science* (7 June 1991) 252:1374-1375 .

Does all of this seem a bit FAR OUT? Indeed, it is. There are several hurdles that need to be overcome, including the following:

1. Efficient cutting chemicals need to be designed for attachment to the custom designed synthetic oligonucleotides that home in on specific (target) nucleotide sequences in double stranded DNA.

2. Methods need to be devised for administering triplex DNA to patients so that this novel drug actually reaches target cells and penetrates the nucleus, where it must act in order to be effective.

Why not think about triplex DNA for a while? Draw out a double-stranded helix, such as the one shown in Figure 3-1 of *EMB*. Then add a third strand that base-pairs with either of the two original strands.

How about designing a method for "piggybacking" an oligonucleotide into the nucleus of a living cell? For example, how would you attach it to a carrier molecule that normally moves from the cytoplasm into the nucleus. Any ideas? How about the DNA replication enzymes, the transcription enzymes, or histones?

Far Out!

Study Questions

1. The base composition of a dsDNA molecule has been determined; the [T]:[G] ratio was found to be 0.4. Given this information, what percentage of bases must be A?

2. A new virus has been discovered, and the genome has been determined to be 22% A, 30% T, 40% C, and 25% G. What does this tell you about the genome of this virus?

3. The DNA molecules below (represented by only one of the two strands) are going to be denatured and then allowed to reanneal. Which of the two strands is *least* likely to reanneal to re-form the original structure? What would prevent that molecule from doing so?

   a. ATATGGTATATATAGGAT   b. GCCTATACGATGTCAGCA

4. Which of the two molecules shown in Question 3 would have the highest T_m? Why?

5. DNA, from phage SP01 that has been labeled with carbon 14 in both strands, is mixed with an equal amount of "normal" SP01 DNA. The mixture is then denatured and allowed to reanneal. What will be the labeling pattern of the resultant DNA molecules (i.e., how will the molecules in the mixture differ with respect to labeling after the procedure is performed)?

6. The same procedure as that used in Question 5 is performed using labeled SP01 and unlabeled T4 DNA. How will the results differ?

7. If base stacking occurs only between bases on the same DNA strand, how can this phenomenon contribute to the stability of double-stranded DNA?

8. Offer an explanation for why the two strands in a DNA molecule are antiparallel, rather than parallel.

9. How many turns are there in a double-stranded B-DNA molecule that is 225 base pairs long? A-DNA?

10. DNA from bacterium A is labeled with carbon-14 and hybridized with an equimolar amount of unlabeled DNA from bacterium B. Fifteen percent of the *total* amount of hybridized DNA has a hybrid density (i.e., one labeled strand and one unlabeled strand). What fraction of base sequences are common to the two species?

11. Why does salt concentration affect the melting temperature of DNA?

12. What are some conditions that facilitate DNA renaturation?

13. A sample of ds DNA is 23% G. What percentage is T?

14. A single-stranded RNA virus contains 22% C. Is it possible to predict the percentage A in this nucleic acid?

15. A linear DNA molecule is hybridized with a molecule that is perfectly complementary, except that the central 10% of the molecule has been deleted.
    a. What is the structure of the heteroduplex that would be seen?
    b. What structure would be seen if the missing 10% of the second molecule were replaced by nonhomologous DNA of the same length?

16. Outline the steps in the dideoxy method of DNA sequencing. Why is it necessary to incorporate a radioactive tracer in the reaction mixtures?

17. Why is RNA more unstable than DNA during the course of laboratory experimentation?

Answers to Chapter 3 Study Questions

1. 14.3%

2. Part or all of the genome must be single stranded.

3. Molecule (a.) is less likely to re-form the original structure because it can hybridize with itself due to the string of ATs flanked by double Gs in the central region.

4. Molecule (b.) would have the highest T_m, because it has a higher percentage of guanine and cytosine.

5. Approximately 50% of the double-stranded molecules have one labeled and one normal strand. Twenty-five percent will contain two labeled strands, and 25% two unlabeled strands.

6. Since the SPO1 DNA and T4 DNA are not homologous, they will not hybridize. Therefore, half will contain two labeled strands (SPO1 DNA) and half will contain two unlabeled strands (T4 DNA).

7. The bases of one strand are chemically linked to the other strand through hydrogen bonding between bases.

8. Most likely it is the result of the constraints imposed by hydrogen bonding between the bases.

9. B-DNA: 22.5 turns; A-DNA: 20.4 turns

10. 30%

11. Salt in a solution of DNA lessens the repulsive forces between the negatively-charged phosphate backbones.

12. high salt, low temperature, high concentration of DNA.

13. 27%

14. No, because the rule, [A]=[T] and [C]=[G], does not apply to single-stranded molecules.

15. a.

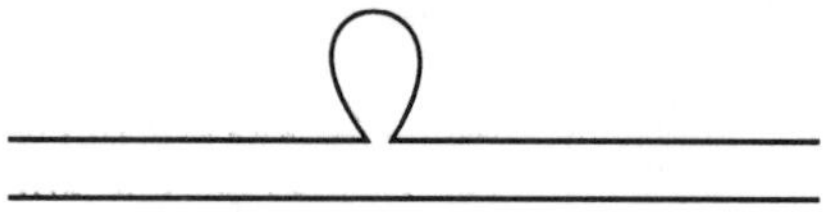

b.

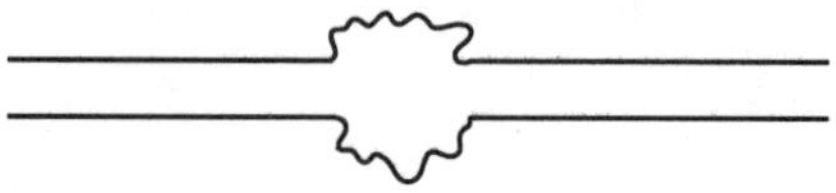

16. Step 1: Divide DNA to be sequenced into four equal portions. Label tubes A, T, G, C.

Step 2: Add DNA polymerase, deoxynucleotides, and one of the four radiolabeled triphosphate dideoxynucleotides (either A, T, G, or C) to each of the four tubes.

Step 3: Let the reactions proceed for the prescribed temperature and time.

Step 4: Run each sample out on one of four lanes of a polyacrylamide gel.

Step 5: Expose the gel to x-ray film.

Step 6: Read the sequence of the newly-synthesized DNA strand from the gel bottom (5' end) to top

(3' end). (Note: samples run from top to bottom).

Step 7: Convert the sequence to the original template sequence.

A radioactive tracer is required to visualize newly-synthesized DNA fragments.

17. 2 reasons:
    1. RNA is inherently less stable than DNA because of the chemical properties of ribose.
    2. RNases that degrade RNA are ubiquitous and are therefore difficult to remove from laboratory glassware, buffer solutions, etc. DNases do not present such a problem.

Concept Map

Let's formulate a concept map using the new information contained in this chapter.

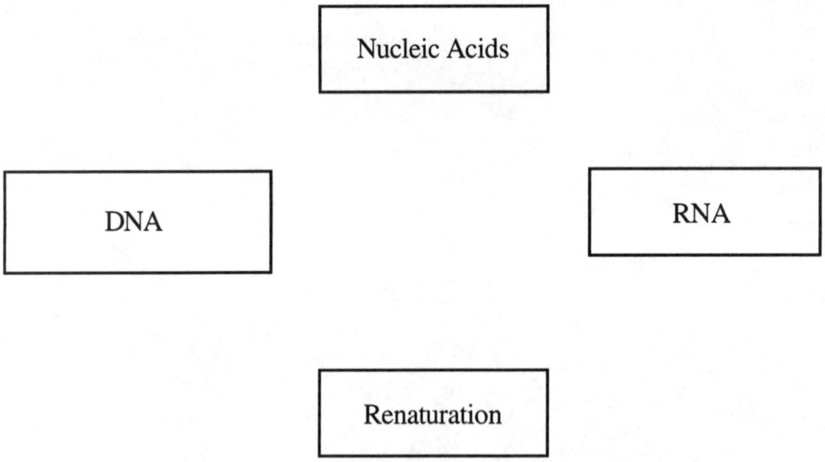

Please use the following terms in your concept map by using either arrows or connecting lines:

- double-stranded helix
- single-stranded
- denaturation
- major groove
- hybridization
- topoisomerase
- sequencing methods
- base pairing
- minor groove
- A, B, Z helices
- supercoiling
- melting temperature
- renaturation
- DNA synthesis

The Molecular Biology Times

DNA Fingerprints to Solve Crimes

Various state officials in collaboration with the Federal Bureau of Investigation are establishing computer data bases that contain the DNA "fingerprints" of persons that have been convicted of violentr crimes. A DNA fingerprint represents an analysis of a single individual's DNA. It is DNA that gives each person his or her individual characteristics (e.g., eye color, hair color, blood type, etc.). Since no two people are exactly alike, except for identical twins, no two DNA fingerprints are identical. This method is also called DNA typing and DNA profiling.

Some legal experts consider this technique to be the single most important technological advance in forensic medicine since conventional fingerprinting was widely adopted. DNA typing is based on the well-established fact that each human being possesses a unique sequence of the four nucleotides that comprise his or her genetic material. This method is predicted to rapidly surpass conventional fingerprinting as the preferred identification tool. With recent technological advances, a typical DNA typing test costs no more than many other routine medical tests. The chances of two humans sharing the same DNA profile has been widely estimated to be one in two hundred million to one in a few billion. Consequently, the DNA fingerprinting method has already been used as evidence in various court trials in the USA and other countries (e.g., England).

This method shows promise of rapidly surpassing conventional fingerprinting as the preferred identification tool.

In order to accumulate a data base for living persons, blood samples must be collected and analyzed, and the DNA type must then be entered into a computer bank. Some versions of the DNA typing test are supersensitive and can be employed to analyze evidence from the scene of a violent crime. Blood (even dried specimens), semen, saliva, hair (with roots attached) and skin scrapings all contain cells, the nuclei of which have DNA. Even minute samples of these cells can be processed in the laboratory using ultramicro methods and then typed. Natural fingerprints are left at the scene of a crime much less frequently than biological specimens. In this way, DNA fingerprinting is more useful than conventional fingerprinting.

The DNA profile method is especially useful for rape cases. Historically, the rate of success in identifying the perpetrator of these crimes has been the lowest of all violent crimes. This method is also useful in victim identification, lost children identity, and paternity disputes. For paternity cases, DNA typing eliminates the need for trials by jury. In fact, attorneys who garner high fees by bringing paternity cases to lengthy, and often highly publicized, jury trials will eventually need to develop another legal specialty.

The DNA typing test is undoubtedly more sensitive than other identification tests and more discriminating than conventional fingerprint methods. Also, the scientific bases of the test is universally accepted as being sound. However, the general public and many courtroom judges have, in many instances, been reluctant to accept DNA typing evidence. Why? Pretrial hearings, at which scientific experts have been requested to testify, have frequently revealed carelessness on the part of the commercial laboratories. It has been discovered that some laboratories are not sufficiently experienced to provide reliable test results.

The method most widely used for routine testing of blood samples has several steps that need to be properly monitored:

extract DNA from
blood cells
↓

cut the DNA into
fragments with
special "cutting enzymes"
(restriction enzymes)

↓

separate the DNA fragments by
means of a sieving system
(agarose gel)

↓

observe the ladder-like banding
pattern of fragments

↓

identify specific nucleotide
sequences in several of
the fragments with radioactive
DNA hybridization
(probe + x-ray film)

↓

compare the banding patterns of 2 samples
on the X-ray film.

↓

are the banding patterns identical
or different?

The cutting enzymes are called "restriction endonucleases," and they are highly specific. They recognize only specific nucleotide sequences. Since no two individuals have the same nucleotide sequence, no two individuals will generate the same size fragments when their respective DNAs are cut with restriction endonucleases. The resulting fragments will exhibit banding patterns on the sieving system that can easily be read by image analysis systems and entered into a computer bank.

A new method can be used to isolate DNA for typing.

A method has been recently developed that eliminates the usual requirement for relatively large samples. This new method relies on a mini-extraction technique that can be used to isolate DNA from a blood stain on a piece of clothing, or even from a few hair follicles. The pure DNA can then be replicated many times using special enzyme technology called "polymerase chain reaction" (PCR). The amplified DNA sample can then be characterized essentially the same way as described earlier.

Scientists are now calling for the establishment of a set of standards or guidelines so that DNA fingerprinting results will be more readily accepted as courtroom evidence. Prosecutors, on the other hand, are attempting to develop a set of legal precedents so that conviction rates will be increased.

Some people believe that the more important issue than standards or legal precedents is that of civil liberties. It is possible that the DNA typing method could become so routine and inexpensive that each newborn baby will be DNA fingerprinted. The data might be transferred to a national, or even international, computer file. This will likely result in questions of confidentiality and due process that may not be very easy to settle.

Science and Society Issues

I. With the advent of relatively inexpensive and reliable methods, DNA-fingerprinting has become relatively routine. Should we all be fingerprinted at birth? Should these records be kept in public files available for inspection?

II. The human genome sequencing project will cost several billion dollars and take 10-20 years. Should government programs should be cut in order to provide the financing for the genome project? If so, which ones?

III. Eventually, it will be possible and relatively inexpensive to sequence portions of an individual's DNA. Should an individual's nucleotide sequence be kept confidential? Should life insurance companies be given free access to this information? Would it be fair if this information were used as the basis for refusing access to low-cost life insurance policies?

Practical Applications Related to Knowledge of DNA Structure

I. Genetically engineered insulin and growth hormone represent some long-term results of the elucidation of DNA structure.

II. DNA fingerprinting as a tool for settling disputes (e.g., paternity cases) or criminal cases (e.g., identifying perpetrators of violent crimes) is becoming widely accepted. Check out the article in the *Molecular Biology Times.*

III. Cancer is now understood and treated as a disease that usually begins as a change in the DNA sequence within a single cell. Carcinogens, including some environmental pollutants, are believed to cause the small alterations in nucleotide sequence that lead to a loss of growth control in cells, resulting in the formation of large tumors.

Solution for "Antiparallel orientation of complementary DNA strands" Exercise!:

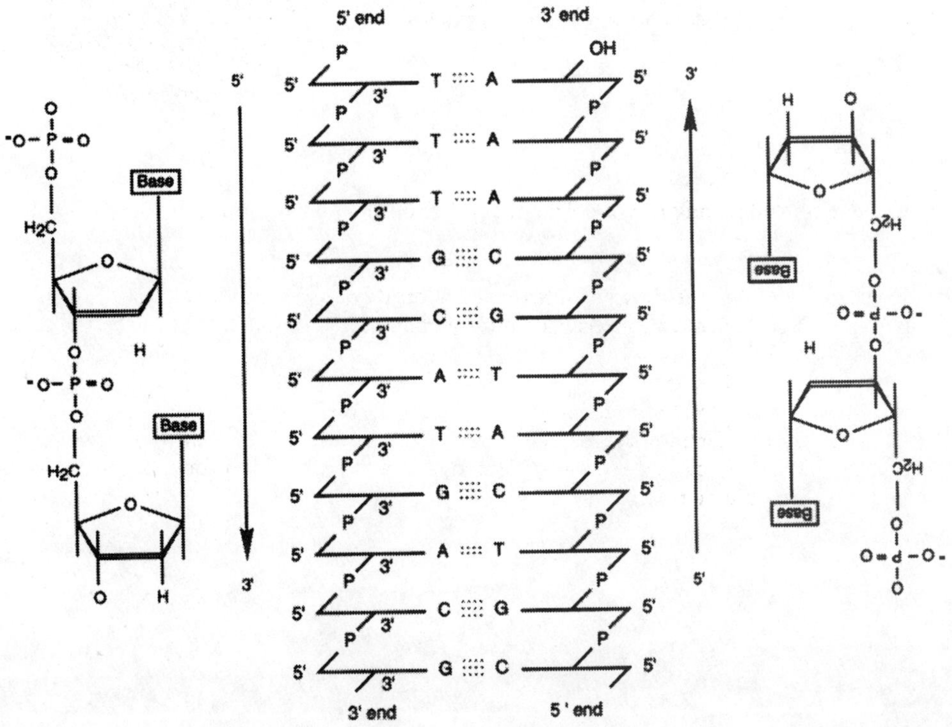

4 The Physical Structure of Protein Molecules

Here's Help

What are the four levels of protein structure?

Primary (1°) structure is the linear order of amino acids in a polypeptide chain. This structure is the key factor in determining the shape and function of a protein. Any change in protein primary structure is usually the result of a change in the base sequence of the DNA molecule. Nevertheless, small changes in primary structure are not always detrimental to protein function. In fact, they may go unnoticed. However, a large change in primary structure usually will inactivate the protein.

Secondary (2°) structure refers to the alpha helices and beta structures in a protein. These "twists and ribbons" (seen in Figure 4-7 in *EMB*) are the result of hydrogen bonding interactions between the amino acids in a single polypeptide chain.

Tertiary (3°) structure refers to the three-dimensional shape of the polypeptide. The tertiary structure of a protein is generated by its primary and secondary structures. Remember, this structure is the combined result of all noncovalent interactions (hydrophobic clustering, ionic bonds, Van der Waals forces, and hydrogen bonds) and covalent interactions (disulfide bonds) that occur among the amino acids of a single polypeptide chain.

Quaternary (4°) structure is the highest level of protein structure. This involves the interaction between two complete proteins. Quaternary structure emerges when two or more polypeptide subunits join, covalently or noncovalently, to form a functional protein. Therefore, a protein that normally exists as a monomer is said to have no quaternary structure.

How about some examples of the importance of 2°, 3°, and 4° levels to protein structure?

Chapter 4 and surrounding chapters include several good examples of the roles that different levels of protein structure play in protein function. One such example illustrates how secondary structure helps determine the extent to which a protein is fibrous or globular. If a protein is composed mainly of either alpha helices or beta structures, it tends to be very fibrous. For example, the major constituent of silk is a protein composed of many beta sheets. Keratin, the major protein component of skin, hair, and feathers, is predominately helical. Proteins that interact with DNA (discussed in Chapter 5), usually have one or more alpha helices that fit into the grooves of the DNA molecule and form weak hydrogen bonds with the base pairs. Another role played by secondary structure is evident in proteins that span the cell membrane. These proteins usually have several alpha helices composed of hydrophobic amino acids that span the membrane. In this case, the alpha helices serve to anchor the protein.

The importance of tertiary structure can be easily seen in enzymes. Enzymes have active sites that are very specific to their appropriate substrates. Active sites are formed as a result of the exact tertiary structures of proteins. When an active site is altered in any way, the activity of the enzyme is either greatly reduced or completely lost.

The discussion of the immunoglobulin G class of antibodies in Chapter 4 provides a good explanation of a protein with quaternary structure. The IgG antibody is formed by the union of four protein molecules; two copies of two different polypeptides. In this case, the union is accomplished by covalent interactions (disulfide bonds). In many enzymes, the tertiary structure includes several hydrophobic amino acids close together on the outside of the molecule. When two or more subunits with these hydrophobic patches on their surfaces occur in close proximity, they will often join at these patches, thereby shielding themselves from water. This is how many large, complex enzymes are formed. However, many other types of interactions also form the basis of quaternary structure.

What is the difference between the lock-and-key and the induced fit models of enzyme activity?

The basis for the lock-and-key model (illustrated in Figure 4-12 of *EMB*) is that the shape of the active site of the enzyme is complementary to the shape of the substrate of that enzyme. Therefore, the only thing that can fit into the active site is the appropriate substrate. All other molecules, including those that are very similar, but not perfectly identical to the natural substrate, are excluded. On the other hand, the induced fit model proposes that the active site of the enzyme changes shape when the correct substrate binds. In this model, binding is due to non-covalent interactions between amino acids and heteroatoms (e.g., zinc, magnesium, sulfur, or iron) of the enzyme, and amino acids of the substrate. The change in shape is what "turns on" the enzyme and initiates the enzyme-substrate reaction. There might be several different molecules in a living cell that can bind to the active site, but only the correct substrate will fully activate the enzyme. However, it is possible that a molecule that closely resembles the correct substrate will cause a low level of activity.

How do the two theories that explain the formation of the enzyme substrate complex differ?

Let's first review, in stepwise fashion, how the lock-and-key mechanism works:

1. The enzyme (E) and substrate (S = molecule to be acted upon by the enzyme) come together. This occurrence is usually the result of a random collision. Thus, the ES (enzyme-substrate) complex is formed.

2. The active site of the enzyme contains appropriately charged amino acid side chains, or other reactive groups (e.g., metal ions). When these side chains or groups react with the substrate, the substrate is transformed to the product. In complex cases, two different substrates are catalyzed by an enzyme at the same time. When this occurs, both substrates simultaneously interact with the enzyme. Then the processed substrates react with each other to form a product.

3. Finally, the finished product is released.

The above mechanism also pertains to the induced fit model. However, when the substrate binds to the enzyme, the enzyme's shape changes slightly. This change in shape improves the ability of the enzyme to carry out a chemical reaction with the substrate. The induced fit model best describes enzymes that consist of multiple subunits. In these enzymes, the subunits are rearranged relative to each other, thereby slightly altering the active site conformation. In addition, the induced fit model provides a better description of ES reactions in which two different substrate molecules react. Binding of the first substrate induces a change in subunit arrangement, which then improves the enzyme's ability to bind to the second substrate.

Why can't we be more specific about the rules that govern folding behavior?

Proteins are remarkably diverse with regard to size, shape, and function. A typical living cell contains between 1,000 and 5,000 different proteins! This means that the relative importance of various forces for folding a polypeptide chain, shown in Figure 4-6 of *EMB*, needs to be considered in the context of a specific protein. Some proteins are rich in hydrophobic amino acids, while others completely lack them. Therefore, it is difficult to state absolute criteria.

How do the relatively strong S-S bonds form?

When the -SH groups from the side chains of two cysteine amino acids come together, a disulfide bridge can form. This happens because the dissolved oxygen in the aqueous environment oxidizes the -SH groups to generate the S-S bonds. These S-S bonds can then stabilize the folded structure.

Let's review the α-helix, shown in Figure 4-4, shall we?

By α-helix, we mean a polypeptide chain that is twisted into a helix. A helix somewhat resembles a spiral staircase. The peptide bonds that comprise the backbone of the protein represent the railing, the hydrogen bonds represent the rungs connecting the rails, and the amino acid side chains that are not involved in hydrogen bonds represent the steps.

In all α-helices, *hydrogen bonding exists in a regular, repeated pattern.* This form of packing is the most energetically favorable and, therefore, relatively stable. If there were more than 3.6 amino acids per turn, then the hydrogen bonds would not be neatly formed and the helix would not be very stable.

How diverse are protein shapes?

Each living cell contains a few thousand different proteins. No two are exactly alike. They differ in size, amino acid sequence, and shape. Of course, these differences mean that proteins carry out diverse functions in a living cell. Some proteins, such as those that comprise cytoskeletons, are usually long and fibrous. Others, like enzymes, usually exist as tight, compact structures. Take, for example, a single polypeptide chain that contains approximately 1,000 amino acids: If it were strung out as a long fiber, then it could measure as much as 1,000 nm (1.0 μm). If the same 1,000 amino acid protein were tightly folded on itself, then it could occupy a small spherical volume less than 10 nm in diameter!

Key Terms

Alpha helix

Beta structure

Lock and key model

Induced fit model

Parallel and antiparallel

Antibody

Antigen

1°, 2°, 3°, and 4° protein structure

H chain

L chain

Constant region

Variable region

Active site

For More Information . . .

Where can I easily find additional explanations of the topics covered in *EMB*'s Chapter 4?

For additional diagrams of peptide bonds and amino acids, look at pp. 56 - 57 in *Molecular Biology of the Cell*, 3rd edition, by Alberts *et. al.*

For help with alpha helices and beta structures, consult pp. 50 - 52 of *Molecular Biology of the Gene*, 4th edition, by James Watson *et. al.* Pages 156 - 158 of the same book provide an explanation of enzyme action.

On pp. 111 - 119 (especially Figure 3-35) of *Molecular Biology of the Cell*, 3rd edition, protein structure is described.

The Wise Owl Says

"In conversation, you can get away with all kinds of vagueness and nonsense, often without even realizing it. But there's something about putting your thoughts on paper that forces you to get down to specifics. That way, it's harder to deceive yourself or anybody else."

Lee Iacocca

P.S.: Could this be one of the reasons molecular biologists, so often when discussing their ideas, sketch them out as diagrams?

Exercise

I. Due to the initial folding of proteins, covalent disulfide bonds form between side chains of cysteines that come within close proximity of one another. Modify the following diagram to illustrate the formation of disulfide bonds.

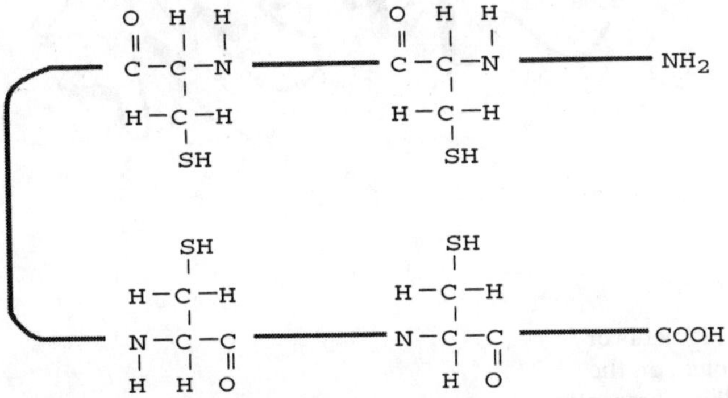

II. Proline is actually an imino acid, rather than an amino acid. The nitrogen atom of the amino group is part of the side chain group.

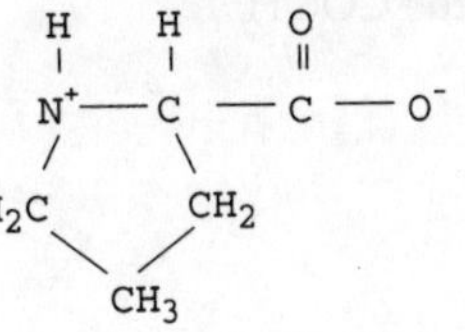

Please draw the chemical structure of a pentapeptide that incorporates proline as the center amino acid.

Notice how the proline causes the backbone structure of the protein to bend. Can you offer a possible explanation for why a proline residue might be found in a protein?

P.S.: The chemical structure is given at the end of this chapter.

III. Illustrate, using simple diagrams, possible hypothetical structures for proteins with the following characteristics. Incorporate into your illustrations all the basic characteristics listed at the left. An example is provided.

Key Characteristics	*Hypothetical Structure*
Example: This protein is a monomer with a cluster of 5 α-helices on the outside, toward the COOH terminus, and 4 β-sheets on the inside, in the middle of the amino acid sequence.	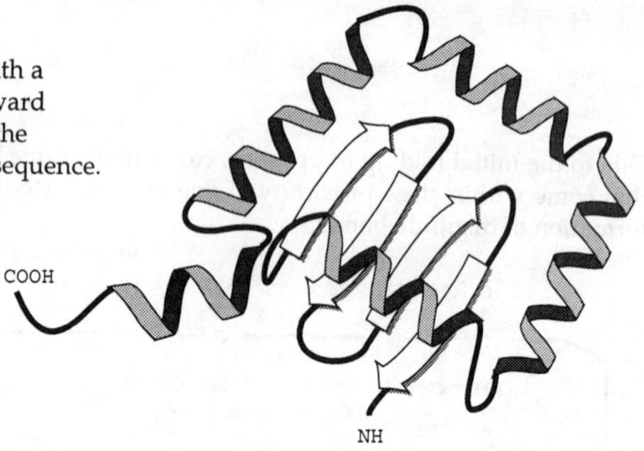
This protein is a monomer. Two segments of β-sheets comprise 30% of the protein on the inside. Three segments of α-helix, representing 20% of the protein, reside on the outside. Large amounts of phenylalanine, tryptophan, leucine, and isoleucine exist at the COOH and NH_2 termini.	
This protein is a trimer, consisting of identical subunits. Approximately 1/3 of the amino acids are glycine. There are also several prolines. Each monomer consists 100% of α-helices. The monomers are wrapped around one another to form a helical rod.	

Tough Nuts

I. Why are the specific primary structures associated with naturally occurring proteins the way they are? That is, if we were to design a specific enzyme from scratch, using a computer graphics program, would we end up with primary, secondary, tertiary, and quaternary structures that resemble the naturally occurring form of that enzyme?

This is a difficult question to answer. That's because the proteins that presently exist in nature are the products of evolution. They evolved to their present structure through modification from preexisting proteins. Information for protein structure is, of course, built into the nucleotide sequence of genes. However, the genes undergo various modifications; including mutation, duplication, deletion, and recombination. It is these random modifications that generate changes in protein structure.

Furthermore, proteins that carry out different, but closely related, functions most likely evolved from a single progenitor protein that underwent evolutionary change in at least two directions. For example, the actin protein found in the microfilaments of a cell's cytoskeleton is different from but closely related to the contractile actin of a muscle cell. Both proteins probably evolved from the same ancestral protein. However, the former evolved towards rigidity and the latter towards contractility. The progenitor, or ancestral, protein probably duplicated one or more times while mutations, deletions, and recombinations occurred in one or more of the duplicate copies. Thus, present day organisms contain proteins that twist, turn, fold, and aggregate the way they do because they evolved from an ancestral protein that exhibited similar twists, turns, and folds.

Now, let's sit down at a computer terminal, with our vast knowledge of protein structure, and design a protein that has a specific function. For example, an enzyme that splits carbohydrate molecules. Would we design a better enzyme than the naturally occurring version? If we could, we might be able to design better enzymes for commercial applications such as laundry detergents (e.g., for "dissolving" dirt) and food production (e.g., for fermenting beer)!

II. Protein folding represents an important research area in contemporary molecular biology. Originally, it was believed that the primary structure of a protein dictated, in a straightforward fashion, a series of intermediate folding steps, *each of which contributed to the final folded structure.*

Recently, however, the scenario of protein folding has become more complex. It now appears that disulfide bonds often play a key role in folding. Yet some of the disulfide bonds present in the intermediate steps in the folding process are not present in the fully folded, biologically functional protein. This observation came as a big surprise to protein chemists! Therefore, protein folding appears to represent a highly cooperative process in which various regions of the primary structure contribute, *in a transient manner*, to the emergence of the fully folded protein.

The key to understanding the complexities of the folding process is to isolate and characterize the intermediates. That, however, is a gigantic task. Some of the intermediates last for such a brief time period that they do not accumulate in the reaction mixture in sufficiently high concentrations to be "trapped," or collected. Other intermediate forms of the folded protein are unstable. Thus, in the process of isolating them they change structure. Unless the experimenter is extremely careful, "false" intermediates, or artifacts, are liable to be generated during the trapping procedure.

How to crack these tough nuts?

Far Out

Chapter 4 emphasized the importance of folding on protein function. That is, in order for a protein to function at optimal efficiency, it needs to be folded correctly.

Most of the information for folding resides in the amino acid sequence of a protein. That is, primary structure dictates the folding properties of a protein. Proteins twist and turn spontaneously and develop into their native globular or fibrous shapes. Does this sound a bit far fetched? Recently, evidence has accumulated that reveals that in the living cell, "helper proteins" called chaperones act on a newly synthesized protein to guide the folding process.

Chaperones bind to the hydrophobic amino acids that are destined to be folded into the core of the protein, and guide them into position. Proteins can indeed fold on their own, without the aid of chaperones. This fact is well known from studies on pure proteins in test tubes. If a pure protein is unfolded by mild denaturing agents, such as high concentrations of urea, and then transferred to an appropriate salt solution, it will then slowly fold back together. However, in a living cell the tendency to fold independently is exploited by chaperones. They speed up the process so that the protein is largely folded even before its entire polypeptide backbone is completely synthesized.

Now let's use our imagination, shall we? Why not exploit chaperones by modifying them in such a way that once they bind to their target protein they remain tightly bound. This would prevent the folding process from proceeding to completion, thereby inactivating the target protein. What could we use "killer chaperones" for? How about using them to inactivate protein toxins that are produced by pathogenic bacteria? Far out?

Study Questions

1. Which types of bonds permit free rotation in a polypeptide? Which do not?
2. In a globular protein molecule, where would one expect to find a leucine residue? An aspartate?
3. Which atoms of amino acids in a polypeptide are primarily involved in H-bonding?
4. What effect do side chains have in the formation of alpha helices?
5. What would be the conformation of polyarginine at pH 2? pH 13.5? Polyglutamate?
6. What is the difference between parallel and antiparallel β-pleated sheets?
7. What would you surmise about the functions of a protein with large amounts of alpha helices? β-sheets?
8. What level of structure (primary, secondary, etc.) is most affected by a drastic change in pH?
9. How does hydrophobicity facilitate the formation of multisubunit structures?
10. What role do disulfide bonds play in IgG?
11. Draw glycine as it would appear at pHs 2, 7, and 13. Do the same for Glu and Gln.
12. What advantages for physiological function do enzymes that consist of aggregates of identical subunits possess?

Answers to Chapter 4 Study Questions

1. Free rotation occurs about the bonds that join peptide groups to α-carbon atoms. Peptide bonds do not allow free rotation.

2. Leucine has a relatively nonpolar side group, incapable of forming hydrogen bonds. It would therefore most likely be found in the interior of the protein molecule, away from the aqueous exterior. Aspartate has a side group that can form hydrogen bonds; therefore it would likely be found on the exterior.

3. The oxygen of the carboxyl group and the hydrogen of the amino group.

4. Side chains are not involved in the formation of α-helices. However, they may disrupt the configuration if they are large or highly charged (largely dependent on pH).

5. The α-helical structure is completely disrupted at low pH (2). Polyarginine exists as an α-helix when the basic side groups are uncharged; this will occur when the pH is high (13.5). pH has the converse effect on polyglutamate, since glutamine has an acidic side group.

6. The parallel structure is composed of chains aligned in the same direction (C terminus to N terminus, or vice versa), whereas chains aligned in opposite directions (e.g., C to N *and* N to C) comprise the antiparallel structure.

7. The protein probably has a structural rather than enzymatic function in both cases.

8. Secondary, since folding is most drastically affected.

9. Hydrophobic interactions are dependent on the exclusion of water. Groups that can engage in hydrophobic interactions, located on the surface of protein molecules, cause binding of the surfaces when water is excluded.

10. Disulfide bonds join the subunits.

11. The amino and carboxyl groups should bear different charges (or no charge at all) at those pHs.

12. Several possible advantages accrue, including the following: In some instances aggregates can carry out a more complex function involving, perhaps, more than one type of substrate molecule. Some aggregates can interact with multiple subunit molecules, thus providing faster overall catalysis. Other types of aggregates are ideally suited for being localized in a membrane or subcellular organelle.

Concept Map

Let's tie this chapter together into a concept map!

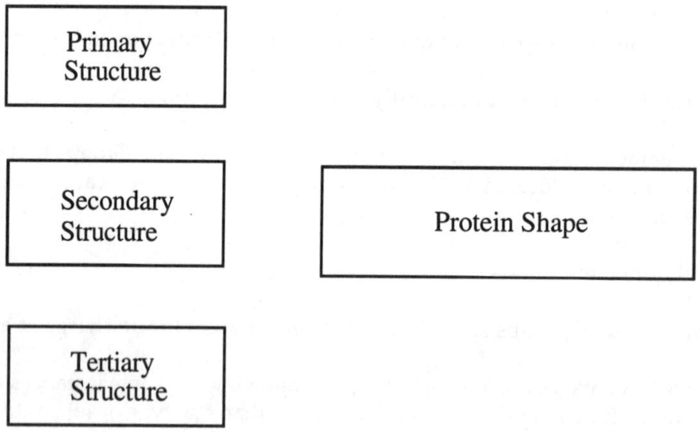

Include the following terms in your network:

folding
α-helix
antibodies
enzyme-substrate complex
disulfide bonds
subunit structure
β-structure
C/V regions

Practical Applications

Cholera, diphtheria, and various other diseases are caused by infectious bacteria that produce protein toxins. These toxins disrupt cellular metabolism. Several of these toxins have unique properties, including the following:

1. They bind to specific receptor molecules on the surface of target cells. For example, the cholera toxin binds to intestine cells.
2. Their subunits can move through cell membranes without disrupting them.
3. Eventually, the toxin kills the cells that it has penetrated.

Knowledge of the tertiary and quaternary structures of protein toxins is very useful to molecular pharmacologists. This knowledge enables them to design vaccines and drugs that attack the toxins. Vaccines accomplish this by enveloping part of the protein, whereas drugs attack by nesting in protein folds.

Cholera toxin, for example, consists of two subunits. The larger subunit is donut-shaped. The shapes of the toxin subunits were recently learned via x-ray crystallography studies. However, before scientists could carry out such studies, the large crystals of pure protein required for x-ray analysis had to be grown. It was a challenge to obtain the large crystals, but once available rapid progress was made.

A hypothesis to explain how cholera toxin acts to debilitate its victim has emerged. First, the donut-shaped subunit anchors to receptors on intestinal cell walls. Then the other, smaller subunit penetrates the cell. Once inside a cell this small protein toxin subunit triggers the expression of genes that cause intestinal cells to secrete the large amounts of fluid associated with diarrhea. Pharmaceutical companies can now begin to formulate "designer drugs." Their aim will be to develop drugs that bind to either subunit in order to short-circuit the toxin's effects.

In addition, knowledge of how the smaller subunit penetrates cells might be useful in designing pharmaceutical agents for delivering other drugs to specific target cells. That is, scientists will use information about the tertiary structure of the small toxin subunit to design carriers for transporting other drugs directly into cells.

Solution for Exercise!:

II. Proline Bend

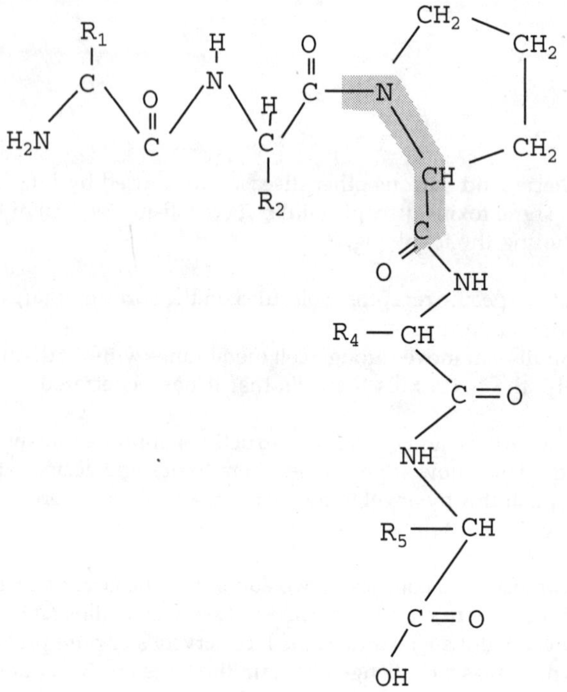

5 Macromolecular Interactions and the Structure of Complex Aggregates

Here's Help

How is eukaryotic DNA compacted into chromatin?

The important factor in the compaction of eukaryotic genetic material is the involvement of histone proteins. First, the octameric disc, comprised of eight histone molecules (two copies of each histone: H2A, H2B, H3, and H4), is completely wrapped *almost* twice by about 146 base pairs of DNA. This forms what is called the **core particle**. Please keep in mind that millions of these core particles organize along a large DNA molecule. The DNA is compact in the octameric disc, but not nearly compact enough for the limited confines of a typical nucleus. The job of histone H1 is to increase compaction even further. It accomplishes this by binding to the free DNA base pairs (about 50-70) between the core particles. This binding causes the DNA to wrap two complete times around the core particle. The correct term to use when referring to the core particle plus the linker DNA and histone H1 is **nucleosome**. Now, this DNA molecule that is complexed with histone proteins is referred to as **chromatin.** The DNA-histone complex, or chromatin, then forms a coil (actually a solenoid) that supercoils and attaches to scaffold proteins.

How do proteins interact with DNA?

Some proteins that interact with DNA do so at **specific** base sequences, others interact with the DNA molecule without showing preference for any particular nucleotide sequence. These latter proteins often have positively-charged amino acids, such as lysine or arginine, on their outer surface. This allows them to interact electrostatically with the negatively-charged sugar-phosphate backbone of the DNA molecule (this is also how histone molecules operate). Proteins that interact with specific base sequences usually have one or more alpha-helices that fit into the major or minor grooves of the DNA molecule. This structure allows the side chains of the amino acids to come in contact with the DNA base pairs and form hydrogen bonds. Additional positively charged amino acids on the protein create electrostatic interactions with the DNA backbone that help stabilize the interaction.

EMB lists several characteristics of protein molecules that interact at specific base sequences. One of the more common, and therefore more important, characteristics is that the protein is usually a dimer that has a pair of symmetrical alpha-helices. These α-helices are positioned such that they interact with the DNA molecule at regions separated by two turns of the helix, or 20 base pairs. (Note: It is very important to know that two turns of the helix is major groove – minor groove – major groove, and *not* major – minor – major – minor – major. Please review Figure 5-13 in *EMB*). If both binding sites do not have the correct base sequence, then binding will not occur.

Another key point is that the base sequence of both binding sites is the same. It might be helpful to look at Figure 5-11 in *EMB*. Binding at sites separated by two complete turns, or at any two sites separated by a factor of 10 base pairs, insures that the protein binds to just one "side" of the helix instead of wrapping around in order to reach both binding sites.

Let's take a eukaryotic chromosome apart, shall we?

Much of our understanding of chromosome structure has been derived from biochemical dissection (e.g., enzymatic digestion, as illustrated in Figure 5-7 of *EMB*) and chemical fractionation (e.g., acid extraction of whole chromosomes). Since histones comprise the predominant type of protein in chromatin, they have been the most thoroughly analyzed. However, it should be kept in mind that a variety of other proteins are present in chromatin, including "DNA binding proteins," such as the *E. coli* Cro protein described in Figure 5-13 of *EMB*. These DNA binding proteins are very diverse in size and structure compared to histones. Histones tend to be small in size and highly basic due to an abundance of arginine and lysine.

Non-histone chromosomal proteins are also present. Nevertheless, they are usually present in so few copies per chromosome, compared to the abundant histones, that they are either lost, ignored, or not recognized in most biochemical dissections. Our biochemical dissection will, therefore, focus on the role histones play in conferring a macromolecular structure to DNA.

Recall that histones play a major role in packing the cell's enormously long DNA molecules into compact units. These units conveniently fit into the nucleus and can be both replicated and expressed (i.e., transcribed into RNA) in an orderly fashion without a lot of tangling.

The fundamental packing unit is the nucleosome. The nucleosome is responsible for generating the "beads-on-a-string" configuration of partially unfolded chromatin that can be visualized in transmission electron micrographs (Figure 5-5 of *EMB*). Once the histones are removed from a chromosome, the superstructure and organization of the DNA in the chromosome is lost:

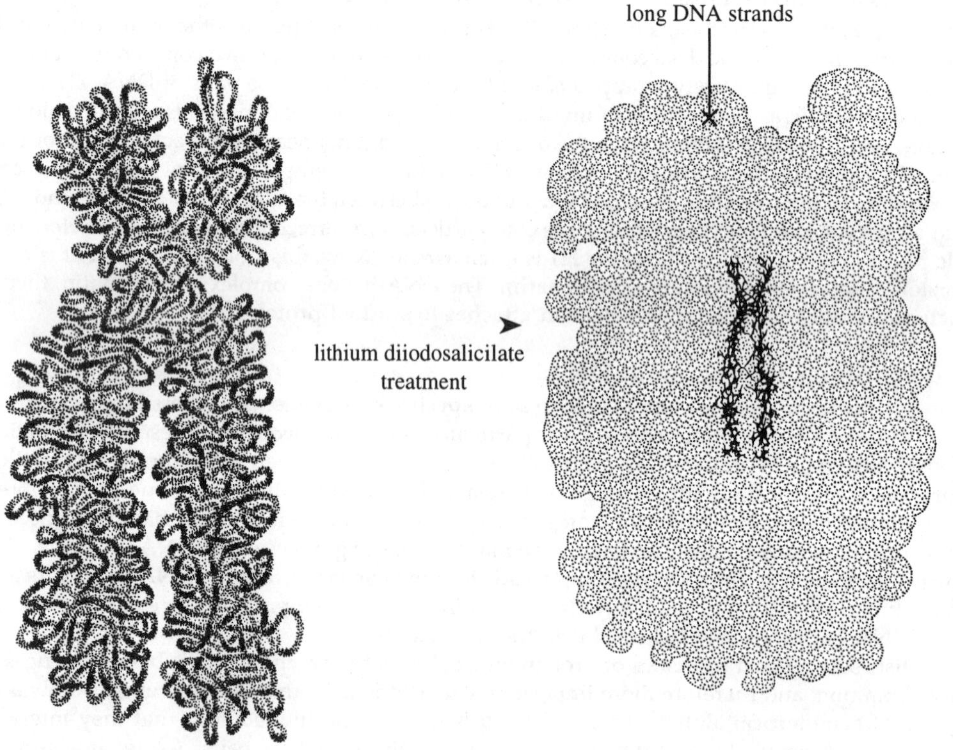

Mitotic chromosome

A non-histone protein scaffold remains. The long strands of DNA are anchored to this central scaffold.

Many molecular biologists speculate that this protein scaffold is responsible for organizing DNA into the familiar mitotic configuration shown in the above (left) diagram.

When chromatin is extensively treated with enzymes that cut DNA, individual nucleosomes are obtained.

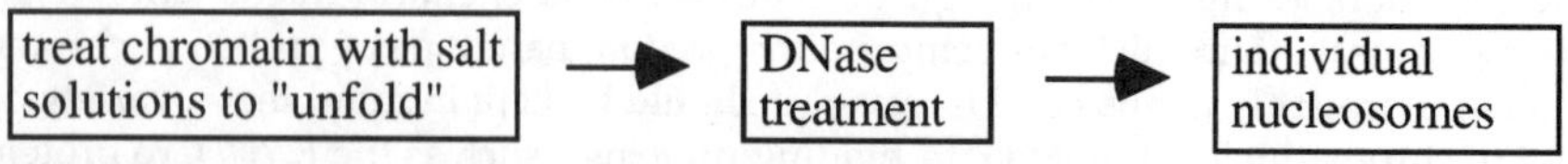

By extracting the histones from individual nucleosomes in a *quantitative* fashion (i.e., the amount of both DNA and individual histones are carefully measured), it can be demonstrated that each nucleosome has the following macromolecular composition:

single nucleosome	low pH (acid) extraction →	approximately 146 bp DNA + 2 molecules of H2A, H2B, H3, and H4

Please note: The DNase treatment above has "trimmed" the "linker DNA" off the nucleosome. This linker DNA is approximately 50 bp long. It is to this so-called linker DNA to which yet another histone protein (H1) binds. Linker DNA is not a special kind of DNA. Rather, the term "linker" is simply used to designate that region of DNA between core particles.

What exactly is the fluid mosaic model?

A biological membrane is a "two-dimensional fluid" composed of two layers of phospholipid molecules. It is called a two-dimensional fluid because, unlike a conventional fluid where the component molecules are able to freely diffuse in every direction, the component molecules of a phospholipid bilayer are only able to diffuse laterally. That is, they can move east or west, but not up or down (north or south). This permits the two layers (inner and outer) of a bilayer to have very different compositions. Thus, a biological membrane is often described as being "asymmetrical".

The fluidity of a eukaryotic membrane is determined by two factors: the types of phospholipids and the amount of cholesterol contained in the membrane. A cholesterol molecule is able to interact with several phospholipid molecules and partially immobilize them, thereby increasing the viscosity and stability of the membrane. Also, a membrane will be less fluid if it is mainly composed of phospholipid molecules with saturated hydrocarbon "tails." Since these tails have no double bonds, they are long and straight. This enables them to come in contact with each other at more points than if they contained double bonds and were "bent." Both the strength of the Van der Waals interactions (as described in Chapter 2 of *EMB*) and the viscosity of the membrane are increased by this configuration. It is important to note that cholesterol is a component only of eukaryotic membranes, not of prokaryotic membranes.

A membrane is not just composed of phospholipids. Many proteins are embedded in it (integral membrane proteins) or attached to its surface (peripheral membrane proteins). These proteins, like the phospholipid molecules, are not able to "flip-flop" from the inner to outer layer and vice versa, but can move only laterally. Transmembrane proteins (which span a membrane in the north-south direction) are actually rare. They are usually anchored to one spot by the cytoskeleton, the cell's complex network of microfilaments and microtubules.

Why do we think membranes are *fluid*?

First, it is highly probable that the bimolecular layer structure shown in Figure 5-17 of *EMB* is the "universal" membrane structure. X-ray diffraction analyses of a variety of membranes from diverse organisms have generated the same type of data: closely packed phosphorylated heads on the surface and fatty acid tails extending internally. This structure lends itself to "fluidity."

Second, a method developed by physical chemists has demonstrated that the phospholipid molecules are constantly moving laterally. "Spin labeling," a method in which the surface portion of the lipid is "labeled" or "tagged" and the generated signal is monitored by electron spin spectroscopy, has revealed that the phospholipids move sideways, but never from one surface to the other. Because the lipid bilayer structure is in a constant state of flux in the living cell, it is called "fluid." It's as simple as that!

What is the significance of the cytoskeleton shown in Figure 5-19?

The cytoplasm (and to a lesser extent the nucleus) of eukaryotic (especially animal) cells contains a 3-dimensional network of fibers, filaments, and tubules which criss-cross the interior of the cell. Those cytoskeletal elements are connected to various components of the cell, including the inside of the plasma membrane (actually, to proteins embedded in the cell membrane), various organelles, and other cytoskeleton components.

Thus, by serving as an anchoring place for macromolecules and organelles, the cytoskeleton provides the cell with an apparatus for organizing its interior! As well, the cytoskeleton gives the cell an overall form and shape. As the cell's shape changes (e.g., during cytokinesis), its cytoskeleton changes (and vice versa).

Key Terms

Nucleoid

Scaffold

Histones

Chromatin

Nucleosome

Core particle

Linker DNA

Non-histone chromosomal proteins

Membrane bilayer

Integral membrane protein

Peripheral membrane protein

Fluid mosaic model

Cytoskeleton

For More Information . . .

Do you want additional explanations on these topics?

More information on histones can be found on pp. 677-678 of Watson *et. al.*, *Molecular Biology of the Gene*, 4th edition. The diagram on p. 678 is especially helpful.

In that same book, on pp. 680-683, there is an excellent explanation of the compaction of the eukaryotic chromosome. You might also want to look on p. 354 of Alberts *et. al.*, *Molecular Biology of the Cell*, 3rd edition, for yet another explanation.

If you felt that *EMB* did not explain the subject of membranes in enough detail, please read pp. 476-484 in *Molecular Biology of the Cell* , 3rd edition.

The Wise Owl Says

Most scientific fields accumulate more information that eventually proves to be either irrelevant or wrong than good information that endures indefinitely, despite the efforts of well-intentioned researchers!

P.S.: A case in point is the structure of plasma membranes. The fluid mosaic model illustrated in Figure 5-17 of *EMB* is relatively recent. Earlier, large amounts of data were collected that portrayed the plasma membrane as a static, rather rigid structure. It was the application of physical and chemical techniques, such as x-ray diffraction, spin-labeling, and the use of cell biology methods (e.g., surface labeling), that revolutionized models of biological membranes.

Tough Nuts

I. Although it is fully certain that eukaryotic DNA is normally complexed with the four major histones, as illustrated in Figure 5-8 of *EMB*, several important questions remain unanswered. For example:

1. Is the nucleosome structure the only macromolecular complex that DNA and histone proteins form?
2. In addition to the four major histones, many cells, including those that comprise embryos, contain a dozen or more variants of the major histones. What are the functions of these variants?
3. What are the dynamics of the nucleosome structure during DNA replication? During RNA synthesis? Does the nucleosome gradually unfold? Does it quickly reassemble?

These questions remain unanswered due to the lack of appropriate experimental methods for directly observing the nucleosome in its natural environment. That is, nucleosomes packed into chromosomes of a *living* cell cannot be studied directly. Most of what has been discovered comes from transmission electron microscope examination of nonliving material and *in-vitro* reconstruction experiments that employ partially purified components. Neither approach allows for direct observation. As a result, our information on nucleosome function is incomplete.

II. The makeup of biological membranes appears, at a quick glance, remarkably simple: Lipid bilayer enclosure with various proteins attached to it, or embedded in it. Is it really that simple? Yes, and no. Yes, in the sense that most membranes in a cell, including those that comprise the endoplasmic reticulum (a network of membranes within the cell cytoplasm) and the plasma membrane (the cell's surface), consist of structures similar to those depicted in Figure 5-17 of *EMB*.

However, the lipid composition varies from one type of cell to another. The major lipids, including cholesterol, phosphatidylcholine, and sphingomyelin, vary substantially in concentration among the membranes found in most cells. Also, the types of proteins (e.g., enzymes) that are associated with any given membrane vary considerably in order to reflect the function of the cell. Actually, it is the other way around: A cell's function is largely established by the proteins that are contained in its membranes!

What makes the composition of a membrane difficult to fully understand is its dynamic (fluid) character. It is always changing. Also, the identification of "minor" components is frequently difficult. For instance, some proteins are present only in low concentrations and others are only loosely attached and therefore easily lost during membrane isolation. Yet, such proteins may play key roles in defining the function of a particular membrane.

III. For almost a decade many molecular biologists have been proposing that the nucleus of a living cell contains a complex network of interconnected protein filaments. These filaments serve as a support system, or skeleton, on which to anchor chromosomes, nucleoli, and other nuclear components. This filamentous skeleton also serves to give the nucleus its shape. Without such a nuclear matrix system, some molecular biologists believe that chaos, disarray, and perhaps even collapse would occur. Furthermore, some researchers believe that DNA synthesis, RNA synthesis, mitosis, and all the other activities that normally occur within the nucleus are spatially organized in various regions of the nuclear matrix. In brief, it is believed by several scientists that a nucleus functions like a well-organized assembly line, with the skeleton playing a key role in maintaining the orderly distribution and movement of various components.

However, obtaining strong evidence to support (no pun intended!) the nuclear skeleton concept has proven to be very difficult indeed. Here is a sampling of some of the obstacles:

1. It is not known what type of proteins and membranous structures might comprise a nuclear matrix.
2. It is unclear whether the high-power photographic images of the nucleus that purport to demonstrate a matrix can be believed. After all, in order to examine a nucleus with either high-power light microscopy, or the transmission electron microscope, it is necessary to "fix" (embalm) the specimen in strong acids, formaldehyde, or similar solutions that are likely to distort, or change, the normal properties of the living cell.
3. It is difficult to determine the exact site of localization, or anchoring, of specific mRNAs, or specific chromosomes, from one nucleus to another. Hence, it is uncertain whether a regular pattern of distribution exists in all nuclei, or whether each nucleus has its own, individual matrix and macromolecule arrangement pattern.

Can you think of some ways to crack these tough nuts?

Far Out

Using our knowledge of macromolecular complexes, including nucleic acid and protein interactions and plasma membranes, let's permit our imagination to run wild. No holding back!

Why not construct "minicells." That is, first isolate the genes for the production of a specific desired product, say human insulin (a protein), or any of a variety of antibiotics. Complex these genes with the appropriate proteins and package them in artificial plasma membranes. Then supply the minicells with the appropriate raw materials and let them go to work. These minicells might have an advantage over conventional "genetically engineered" microorganisms (see Chapter 15 of *EMB*), for they could perhaps be designed for doing only one thing: producing a single product. No need to grow or metabolize.

These minicells might function like microscopic robots. A whole factory or production facility could conceivably be organized in a single small room. In much the same way that automobiles are produced largely by robots in state-of-the-art assembly factories, biological products might be produced with microscopic robots. Far out?

P.S.: What would we want to call our minicells? Here are a few suggestions to consider, "mini-mimic cells" (MMCs), "microrobot cells" (MRCs), and "artificial biological processors" (ABPs).

Study Questions

1. Given a protein that does not bind to ssDNA, only to dsDNA, in low salt concentrations. Furthermore, this protein does not bind to any specific sequence. Why the effect of salt concentration? Where must the binding site be located?

2. Given a protein that binds only to dsDNA and binds regardless of salt concentration. Also, the protein does not bind to any specific sequence. Where must this binding site be located?

3. Given a DNA-binding protein that is a dimer. Each monomer has only one binding site. What can you infer about the location, on the protein, of that site?

4. There have been no observations of mutations that give rise to altered histones. Why not?

5. Most membranes have an unequal distribution of different types of phospholipids (the inner layer will have more of one kind, while the outer will have more of another). What factors act to maintain this asymmetry?

6. Why is it beneficial to cells that transmembrane proteins don't flip-flop?

7. What is the purpose of histone H1?

8. What is the significance of histones being rich in lysine and arginine?

9. The amino acid sequences of the various histones are virtually identical among all organisms. Why do you suppose that is so?

10. Discuss both the active and passive means that allow molecules to pass through membranes.

11. Regulatory proteins bind to specific base sequences. What interactions between DNA and protein specify the binding? Describe a test that will determine the location of those binding sites.

12. Compare prokaryotes and eukaryotes in terms of how they condense their genetic material. Why is this condensation necessary? Give at least two reasons.

13. Would you expect the compaction of eukaryotic DNA to vary at different stages in the life cycle of a cell? Explain your answer.

Answers to Chapter 5 Study Questions

1. The binding site must be on the negatively-charged backbone of the DNA molecule since binding occurs only at low salt concentrations. High salt lessens the negative charge of the backbone thereby decreasing the strength of the binding.

2. Since binding is independent of salt concentration, the negative charge of the backbone is not involved. Most likely the binding occurs in the major or minor groove by bonds unaffected by salt concentration (e.g., hydrogen bonding).

3. The binding site on each monomer is probably comprised of α-helices that bind to the major groove of dsDNA. The binding sites of the dimer are probably related by two-fold symmetry to increase binding capabilities in the major groove (e.g., Figure 5-13 of *EMB*).

4. Histones are functionally constrained and, therefore, highly conserved. That is, they probably have evolved the optimal structure for carrying out their highly specialized function.

5. The extracellular (outside), fluid environment (salt concentration, pH, etc.) is slightly different from the intracellular (inner) environment. Hence, the phospholipids that are exposed to the outer surface are best suited for interaction with the outside environment, and vice versa.

6. The region of an integral protein occurring at a cell's surface and the region on the inside of a cell are designed for their respective environments. Flip-flopping would expose these regions to environments in which they are not designed to function.

7. The H1 histone serves to connect the octamer and the linker DNA, thereby drawing them closer together and making the nucleosome a more compact structure.

8. Lysine and argine are positively charged amino acids that bind to the negatively charged DNA backbone.

9. Histones are functionally constrained and therefore highly conserved. That is, since they serve the same function in all organisms, their amino acid sequences are wholly the same in all organisms.

10. Active means include the chemical modification of either membrane components (e.g., channels) or the transported molecules themselves. They are usually energy-requiring processes. Passive means include diffusion directly across the lipid bilayer and diffusion through channels. These processes do not require energy.

11. Various noncovalent interactions, including hydrogen bonding, ionic bonding, and hydrophobic interactions. A test that will determine the location of these binding sites is DNA footprinting: DNA and protein are allowed to bind, the complex is digested with DNase, and then the regions which were not cut by the DNase are determined. These regions represent those that are bound by the protein and therefore are protected from the action of DNase.

12. Prokaryotes condense their genetic material by superhelicity and by attaching it to scaffold protein: a relatively loose packaging. Eukaryotes condense the genome much more compactly via interactions with histone proteins. Condensation is necessary for the genome to fit into the nucleus. Also, condensation protects DNA from DNase digestion, breakage, and other damage.

13. Yes, it is most densely packed during cell divisions. This makes sense since chromosomes must separate. The genetic material is the least densely packed during the G1 and S phases, during which time the synthesis of messenger RNA and replication occurs, respectively.

Concept Map

The relationships explained in Chapter 5 of *EMB* will be better defined by preparing a concept map.

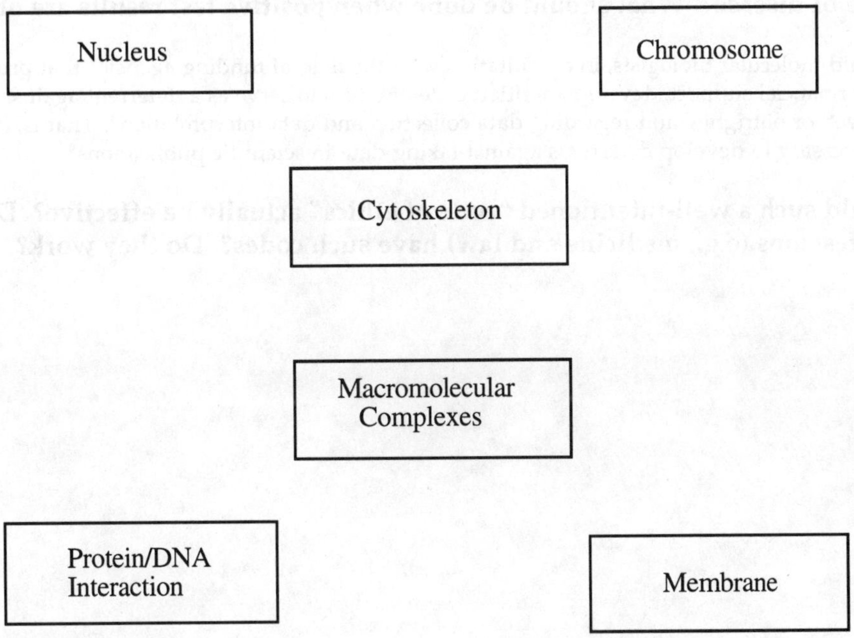

Please include the following terms in your concept diagram by connecting them with lines or arrows:

DNase digestion	histones
scaffold	DNA
nucleosome	fluid mosaic model
Cro protein	actin filaments
chromatin	microtubules
bilayer	intermediate filaments

Science and Society Issues

I. Some genetic diseases, such as cystic fibrosis, appear to involve alterations in most, if not all, membranes in the affected individual. That is, the membrane structure illustrated by Figure 5-17 of *EMB*, especially the protein components, is thought to be defective. Since so many membranes are defective, a complete cure of all of the patient's symptoms is unlikely to be developed.

Should prenatal cell samples of fetuses be routinely tested for indications of this type of disease? What should be done when positive test results are obtained?

II. Should molecular biologists, in consultation with the federal funding agencies that provide most of their financial support, develop a written code of ethics to serve as a deterrent against cases of abuse, neglect, or outright fraud regarding data collection and data interpretation? That is, do you think it is necessary to develop deterrents against faking data in scientific publications?

Could such a well-intentioned "code of ethics" actually be effective? Do other professions (e.g., medicine and law) have such codes? Do they work?

Practical Applications Related to Understanding Macromolecular Complexes

Molecular medicine hopes to be able to develop agents that inactivate specific membrane receptor proteins`. Such agents might be useful therapeutic approaches to controlling drug addiction. For example, caffeine, cocaine, and heroin act by first binding to specific receptor proteins that reside in the plasma membrane of many cells, including those of the human nervous system. This type of drug lends itself to addiction as well as, hopefully, to this type of "molecular therapy."

6 The Genetic Material

Here's Help

Let's review the experiments discussed in Chapter 6.

Four monumental experiments are discussed in Chapter 6. Together, these four experiments conclusively demonstrated that DNA is the genetic material. This is true in virtually all organisms, the notable exception being RNA viruses. The first experiment discussed is the transformation experiment, performed in 1928 by Fred Griffith. (See Figure 6-2 of *EMB*.) Two strains of the bacterium *Streptococcus pneumoniae* were used for this experiment. One had a polysaccharide coat around the cell wall, the other did not. Griffith discovered that when the strain that lacked the coat was in the presence of the other strain, it acquired the ability to grow a coat. This experiment did not prove that DNA was the genetic material. In fact, it didn't even suggest it. This experiment simply proved that some, then-as-yet uncharacterized, genetic material existed. That is, a "transforming principle" capable of carrying genetic information from one cell to another was discovered.

As time passed, technology improved. Fifteen years after Griffith's transformation experiment, Avery, MacLeod, and McCarty were able to improve the experiment. (See Figure 6-4 of *EMB*.) This time it was shown that DNA comprised at least part of the genetic material. However, they were not able to completely purify the DNA and rid it of impurities that could have confused the interpretation. Thus, they were not successful in proving, beyond a shadow of a doubt, that DNA alone was the transforming principle.

The next experiment discussed is the Hershey-Chase, or blender, experiment. This experiment demonstrated that DNA alone is the carrier of genetic information. Since viruses consist mainly of DNA and protein and they reproduce by injecting their genetic material into a host cell, Hershey and Chase labeled the DNA with ^{32}P and the protein with ^{35}S. They then checked to see whether the DNA or the protein was injected into the cells. As explained in Figure 6-6 of *EMB*, the bacteria acquired the ^{32}P and not the ^{35}S during the infection process.

The last experiment discussed, the transfer experiment, simply supported the findings of the Hershey-Chase experiment by showing that viral DNA could infect naked bacterial protoplasts.

The transformation and blender experiments were both done using prokaryotes. What about similar experiments using eukaryotes (e.g., mammals)?

Similar experiments, such as the transformation experiment, are now done routinely with cultured mammalian cells. It is possible, for example, to culture mammalian cells that lack a particular enzyme activity and "transform" them by adding purified DNA that contains the missing gene directly to the cells. This procedure is analogous to the procedure originally used for bacteria (e.g., Figure 6-3 of *EMB*).

In addition, "transgenic" experiments that involve directly injecting a pure gene (DNA) into a mammalian egg are routinely carried out in several laboratories. The results are, of course, "transformed" animals (such as mice, sheep, etc.). You will eventually read about this new technology in Chapter 15 of *EMB*.

Why is a template molecule such as DNA required to transmit information?

Consider proteins as being the foot soldiers of the living cell. Virtually all the properties of a cell, including its size, behavior, and function, reflect its protein composition. As cells divide, they must reproduce themselves in order to "stock" the daughter cell. But how can a protein (e.g., an enzyme) make a copy of itself? Or, for that matter, a copy of another protein? Here is where the concept of a template comes into the picture. The template DNA, via its intermediate messenger RNA, specifies the sequence of amino acids in various proteins. Since the template DNA can be copied (during DNA replication—see Chapter 7 in *EMB*), the foot soldiers can be indirectly multiplied. Double-stranded DNA has evolved as an excellent template, not only for replicating itself, but for protecting the valuable information that resides in the sequence of the nucleotide bases. Some of the virtuous properties that DNA offers are listed in Chapter 6 of *EMB* under **Properties of the Genetic Material.**

Why is deaminated cytosine repaired in DNA but not in RNA?

The reason is actually very simple and shows how economically cells operate. If cytosine in DNA deaminates to uracil and is not corrected, then all RNA made from that region of DNA will be incorrect and will most likely make defective proteins . Likewise, if a cytosine in RNA is deaminated, all the proteins made from that RNA will most likely be incorrect. The cell has a rather simple mechanism for repair in DNA, but not RNA. The question, "why not?", then naturally arises.

One explanation is, that DNA is not supposed to contain uracil, so if any is present it must be from the deamination of cytosine. Enzymes scan DNA and recognize the odd bases. However, it would be nearly impossible to edit improperly located uracils in RNA. How could an enzyme distinguish a "good" uracil from a "bad" uracil in RNA? RNA already contains lots of uracil!

A more "conceptual" explanation, however, has to do with economics. Hundreds or thousands of mRNAs are made from a single DNA molecule, each resulting in the synthesis of many proteins. So, when a C changes to a U in DNA, every RNA molecule and *every protein molecule,* made from that segment of DNA, will be affected. Obviously, therefore, cytosine deamination needs to be repaired in DNA. Now, let's look at RNA. If thousands of RNA molecules are made from a particular segment of DNA and a few of those acquire "bad" uracils through cytosine deamination, so what? There are still thousands of correct RNAs from which tens of thousands of correct proteins can be made. Furthermore, DNA molecules persist for a long time (much longer than the cell's lifetime), while RNA molecules usually degrade within a few minutes or hours and more are made as needed. Could it be that it's simply not worth the cell's time and "money" (energy) to make the enzymes to repair deaminated cytosine in *RNA*?

Key Terms

Codon

Deamination

Ribonuclease

Chemical alteration

Replication error

Blender experiment

Hershey-Chase experiment

Transformation experiment

Transfer experiment

For More Information . . .

Would you like additional explanations on these topics?

Regarding the experiments covered in this chapter, consult pp. 69-73 of Watson *et. al., Molecular Biology of the Gene,* 4th edition.

For additional help with cytosine deamination, please review Figure 6-33 on p. 245 of the 3rd edition of Alberts *et. al., Molecular Biology of the Cell.*

The Wise Owl Says

Human intuition, although powerful, isn't always sufficient for unraveling the complex molecular mechanisms of the living cell.

The scientific literature on molecular biology provides evidence that some mechanisms are counterintuitive. Here are two especially prominent examples.

1. Since DNA is comprised of only four nucleotide bases (A, G, C, T) the genetic information was initially believed to be stored and transmitted by other molecules, such as proteins (which contain 20 different amino acids). The experiments outlined in Chapter 6, however, resolved this issue.

2. It was assumed that eukaryotic and prokaryotic genes were structurally alike because eukaryotic genes are comprised of nucleotide sequences that are surprisingly similar to bacterial genes. Not so. In 1977, it was discovered that eukaryotic genes contain frequent "interruptions," in the form of nucleotide sequences (called "introns" to be discussed later in Chapter 8 of *EMB*), that are processed out of the mRNA.

Tough Nuts

I. Why did so many biologists pass by DNA when trying to discover the chemical identity of the genetic material? Historians offer several explanations:

a. Chromosomes contain large amounts of RNA and protein, as well as, of course, lots of DNA. The DNA was considered, by many biologists, to function as a skeleton for chromosome structure.

b. Nucleic acids are relatively simple in chemical composition because they contain only 4 bases [A, G, C, T (or U)]. On the other hand, proteins contain up to 20 different amino acids. "Four" was considered to be an insufficient number for generating biological diversity. "Twenty," however, was considered sufficient. Needless to say, the discovery of the triplet genetic code (see Chapter 9 in *EMB*) was not made until several years after DNA was unequivocally recognized as the genetic material.

c. The crude biochemical extraction methods that were first employed for the characterization of nucleic acids yielded small fragments. Substantial degradation of large DNA polymers easily occurs if, during each phase of the isolation, special precaution is not taken to avoid mechanically shearing the polymers into small pieces.

II. What changed the minds of biologists concerning DNA as a candidate molecule for the genetic material? No single experimental result convinced everybody. Instead, cell biologists accepted some discoveries more easily than others, while microbiologists weighted other data more heavily. Ultimately, there was a cumulative effect. So many lines of evidence eventually pointed to DNA that it became difficult to *not* believe that genes were made from DNA. The following lines of evidence accumulated:

a. Improved extraction methods led to the identification of chromosomal DNA as a very long polymer to which RNA and protein were loosely bound.

b. Improved visualization methods, especially the transmission electron microscope, revealed DNA as being a long (and thin) molecule in its natural state.

c. The transforming principle discovered by Griffith (see Figure 6-3 of *EMB*) was demonstrated to be destroyed by treatment with the enzyme DNase. Neither RNase nor proteolytic enzyme treatment were capable of destroying the transforming principle.

d. Careful chemical measurements revealed that diploid cells (e.g., normal body cells) contain twice as much DNA as haploid cells (e.g., sperm cells). This is consistent with the observation that cells that have completed meiosis (e.g., sperm cells) contain half as many chromosomes as normal body cells.

e. It was furthermore observed that the amount of nuclear DNA is constant from cell to cell within an organism. For example, liver, kidney, and pancreas cells all contain the same amount of nuclear (i.e., chromosomal) DNA.

f. Virtually all DNA was demonstrated to reside in a cell's nucleus. Only trace amounts were found in the cell's cytoplasm.

g. Metabolic studies revealed that DNA is normally stable in living cells, whereas RNA and protein are in a constant state of flux. That is, they are continually synthesized and destroyed.

As you can probably imagine, the evidence for DNA as the genetic material became overwhelming. Most molecular biologists agree that the blender experiments (e.g., pp. 103-105 in *EMB*) persuaded even the most skeptical scientists that DNA is the genetic material.

Then, of course, came the interpretation of x-ray diffraction data. It provided the "icing for the cake." This data, when interpreted in terms of the now famous double-helix structure, explained several key features that the previous characterizations of DNA had not dealt with. The double-helix model explained why [A] = [T]; [G] = [C]. It offered a plausible idea for DNA replication; the semi-conservative model (see Chapter 7 in *EMB*). Also, it accommodated RNA as an intermediate in the transfer of information from DNA to protein structure.

III. Although we just made a strong case for DNA as the genetic material, we cannot just write RNA off as a mere "information transfer vehicle." Certain viruses employ RNA as the genetic material! For example, plant viruses, such as the tobacco mosaic virus, were first shown in 1956 to contain an RNA genome. This RNA is single-stranded except during replication, when a double-stranded intermediate is involved. There are yet other viruses, including some mammalian viruses that cause tumors in animals, that synthesize a single-stranded DNA molecule from an RNA template. This single-stranded DNA molecule then serves as a template for the synthesis of a complementary DNA strand. The double-stranded DNA is then inserted into the host's chromosome and, when expressed, it can also cause tumor formation. The AIDS virus, discussed in Chapter 16 of *EMB*, is yet one more example of an RNA-genome virus.

Study Questions

1. In the blender experiment, why were ^{32}P and ^{35}S used as markers instead of some other radioactive elements?

2. How might the blender experiment need to be altered if a phage other than T2 were to be used (i.e., a phage that didn't have a tail or that was engulfed by the bacterium)?

3. Explain how DNA was found to be the genetic material by Avery, MacLeod, and McCarty's experiment.

4. Why is mRNA a valuable intermediate in protein synthesis?

5. *EMB* states that hydrogen bonds are essential for correct transmission of the genetic information to progeny. In your own words, explain why this is so.

6. Why might the 2'-OH group found in ribose, and not deoxyribose, contribute to the instability of RNA?

7. What is meant by bacterial transformation?

8. How are uracil and thymine chemically related?

9. Assume that mutation repair systems are less than 100% efficient and, therefore, help to maintain a mutation rate that is "very low but useful in the long run." What is meant by this statement?

10. Why is it important that uracil not be employed as a base in DNA?

Answers to Chapter 6 Study Questions

1. Phosphorus and sulfur are constituents of DNA and proteins, respectively. Therefore, their radioactive forms can be incorporated into the macromolecules so that only one (e.g., DNA) and not both are labeled.

2. Another means of separating phage from bacterium, that doesn't depend on the relatively weak phage tail, would have to be used. For example, a biochemical fractionation procedure could be one method.

3. The transforming principle, known to transform the rough strain of *Pneumococcus* to the smooth strain, was purified and identified as DNA.

4. RNA is chemically less stable than DNA. Thus, it is quickly degraded, often minutes after it is produced. This affords the cell a sensitive means of regulating protein production, by rapidly replacing the mRNA population.

5. Hydrogen bonding governs the specific base pairing of guanine and cytosine, and thymine and adenine. Thus, linear nucleotide sequences are preserved in the offspring through the synthesis of complementary nucleotide sequences. Hydrogen bonding provides the weak forces for pairing correct bases during DNA replication.

6. In the presence of a hydroxyl ion (OH^-), that could come from a base, the 2'-OH of the ribose sugar increases the rate of hydrolysis of the phosphodiester bond at the 3' oxygen linkage to the ribose, thus cutting the RNA.

7. Bacterial transformation is the production of a new phenotype as a result of the introduction of novel genetic material.

8. Thymine has a methyl group in place of one of the hydrogens found in uracil.

9. Mutation is the source of novel genes on which natural selection acts. This arrangement, mutation followed by natural selection, allows for the evolution of new species.

10. Because deaminated cytosines are indistinguishable from uracils (because they are uracils!), an organism employing both cytosines and uracils in its genome could not identify and, therefore, could not repair deaminated cytosines.

Concept Map

Shall we pause and tie it all together in a concept map?

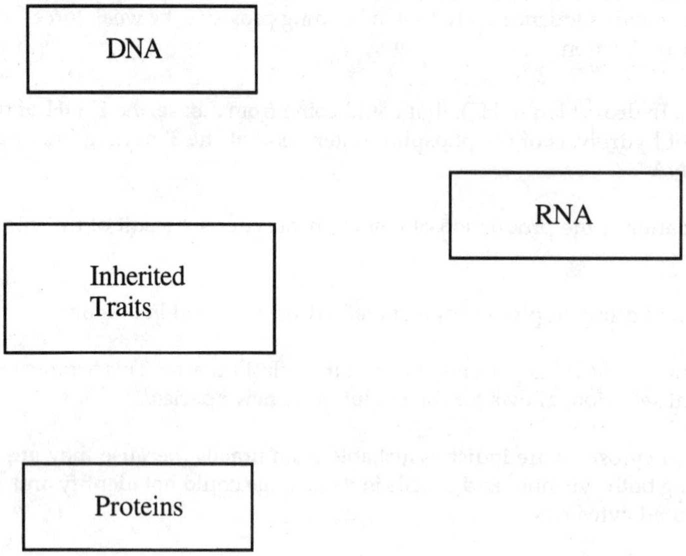

Please add arrows, or connecting lines, that will illustrate logical connections between the above boxes. Please add, also with connecting arrows or lines, the following components:

- Hershey-Chase experiment
- amino acid sequence information
- transformation experiment
- heat-killed bacteria
- DNA identified as transforming principle
- phage T2
- radioactive phosphorous and sulfur
- mutation
- physical stability of DNA
- expression of information
- messenger RNA triplets
- DNA replication
- sugar-phosphate backbone
- RNA as genetic material
- viruses

7 DNA Replication

Here's Help

How important is Chapter 7?

There is a lot of information in this chapter. Most of the material included in earlier chapters has been "building up" in order to prepare you for Chapter 7. All of the material in subsequent chapters will be based upon the information contained in this chapter. Learning this information is critical, since most of the enzymes and processes that are discussed here will reappear in the later chapters.

What is meant by the term "semiconservative replication"?

As you know, both strands of the double-stranded DNA molecule are copied, or replicated, prior to cell division. Then one copy of the replicated molecule goes to each daughter cell during cell division. The DNA replication process yields two new daughter strands of DNA. The two old, parental strands were used as templates for the daughter strands. They are "conserved." When this was first being understood, the following question arose: Are the two new double-stranded molecules composed of one daughter and one parental strand, or do the two parental strands remain together, while the two daughter strands form a "brand new" double-stranded molecule? The latter model is referred to as the "conservative" model of replication, since in this model the original double-stranded molecule is conserved during replication. The former is called the "semiconservative" model. By some rather elegant experiments, involving labeling the newly synthesized DNA with isotopes (not discussed in *EMB*), the semiconservative model was demonstrated to be correct. In other words, after cell division, each daughter cell has a "half-new, half-old" DNA molecule. If you would like to know more about this experiment, see "For More Information."

What do the terms "continuous" and "discontinuous" synthesis mean?

Continuous synthesis means that a daughter strand of DNA is synthesized as a single unit; the replication enzyme starts and keeps going to completion, generating a long, continuous strand. Discontinuous synthesis, on the other hand, occurs when the daughter strand is synthesized in short segments that are later linked together to form a long molecule.

The leading strand in a replication fork (growing in the 5'→3' direction) can be synthesized continuously because there is always a free 3'OH available for DNA polymerase. However, there is not always a free 3'OH available on the lagging strand, so synthesis must be discontinuous. In order for discontinuous synthesis to work properly, the replication machinery must periodically stop and wait for the fork to open up enough to allow the placement of a free 3'-OH of a base-paired oligonucleotide, usually in the form of an RNA primer.

Did you notice the phrase "free 3'-OH" several times in the previous discussion?

Very perceptive! A free 3'-OH must ***always*** be present for DNA synthesis, of any kind, to occur. Of course, a template, the four deoxyribonucleotide 5'-triphosphates, and all the appropriate enzymes must also be present to synthesize DNA. But the presence of a free 3'-OH of a pre-existing DNA strand for attachment of subsequent nucleotides is essential for DNA synthesis. This is what primers are all about. They are simply RNA or DNA molecules that provide a free 3'-OH for the initiation of DNA synthesis. Note, however, that the enzymes involved in RNA synthesis do not have this restriction; this is why primers for DNA polymerase are usually made of RNA. That is, *RNA* polymerase, unlike *DNA* polymerase does not need a primer to initiate synthesis of RNA.

Why does DNA synthesis begin at specific (*ori*) sites, rather than just any place along the DNA molecule?

In order for a *replication bubble* to form, it is of course necessary for double-stranded DNA to separate, into the single strands which comprise the replication bubble shown in Figures 7-6, 7-7, 7-8, 7-24, and 7-25 of *EMB*. Breaking of the hydrogen bonds which hold single-stranded DNA in the double-stranded helical configuration is easiest to accomplish at A-T rich regions, since you may recall that an A-T base pair is less strongly bonded (2 H-bonds) than a G-C base pair (3 H-bonds). Figure 3-2 back in Chapter 3 of *EMB* illustrates that hydrogen bonding.

Thus, regions of genomes have evolved which are A-T rich, and which have nearby base sequences ("9-mers" in Figure 7-7 of *EMB*) which serve as binding sites for proteins which twist DNA, and in the process undo the hydrogen bonds in the A-T rich regions. The resulting "bubble" or "open complex" that forms provides ample opportunity for the various DNA synthesis enzymes to bind and begin copying the two single-stranded templates.

What do *topoisomerases* accomplish for DNA replication?

They relieve the twisting that occurs as helicase unwinds DNA during the replication of a covalently-closed circular DNA molecule. Figure 7-4 of *EMB* illustrates that excess twisting. Thus, these enzymes alleviate the twisting and tangling which would result as the replication bubble gets larger and larger.

How about a concise summary of the conclusion which data from the pulse-chase experiment (Figure 7-17 of *EMB*) generated?

Try this one on for size: Some DNA is replicated as short fragments, which are later joined to form a longer strand!

After reading the text (e.g., Table 7-1), are you still unclear about activity differences between DNA polymerases I and III? Here's help!

DNA Pol I:

a. Polymerization activity in the 5'→3' direction *ONLY*.
b. Exonuclease activity in the 3'→5' (opposite) direction (a.k.a. the "editing function").
c. Exonuclease activity in the 5'→3' direction used to remove lagging strand primers (a.k.a. "nick translation").
d. Strand displacement activity.

DNA Pol III:

a. Polymerization in the 5'→3' direction *ONLY* (like Pol I).
b. Exonuclease in the 3'→5' direction (a.k.a. the "editing function") (like Pol I).
c. No important exonuclease activity in the 3' direction and no strand displacement (except in rolling circle replication).

Here's a step-by-step review of the events at the replication fork:

Please keep in mind that DNA in a living cell usually exists in large covalently closed circles that are twisted (i.e., supercoiled) around themselves. Figures 3-7 and 3-8 in Chapter 3 of *EMB* illustrate this. For replication of such a large circle to be completed without a lot of tangling, the supercoiling must be temporarily relieved. This difficult aspect of DNA replication is often called the "topology" issue. Various topoisomerases (e.g., Figure 7-5 of *EMB*), enzymes that manipulate complex DNA superstructures (including DNA gyrase), accomplish this unfolding and unwinding.

Initiation of the replication process follows the unfolding and unwinding process. Specific nucleotide sequences, usually rich in A-T base pairs, exist in most DNAs for facilitating the initiation process. These sequences, by virtue of being A-T rich, "breathe" easily. That is, the two hydrogen-bonded DNA strands easily come apart, or melt, allowing the polymerizing enzymes to gain access to the template. These specific nucleotide sequences are called "origins of replication," or *ori* (see Figures 7-7 and 7-24 in *EMB*).

Since the two DNA strands are antiparallel (one runs 5'P-3'OH *left to right*, while the other runs 3'OH-5'P *left to right*), it is impossible for a single enzyme system to simply bind to the separated strands and in the same direction synthesize a new daughter strand from each of the parental template strands. This is because DNA polymerase possesses a highly specific catalytic activity. It reacts deoxyribonucleotide triphosphates with the 3'OH of a preexisting strand, or primer. It simply *cannot* catalyze the reaction of the 3'OH of a nucleotide triphosphate with a 5'P of a preexisting primer, alone, without having a primer.

The first illustrations (below and on the following page) show how DNA replication works:

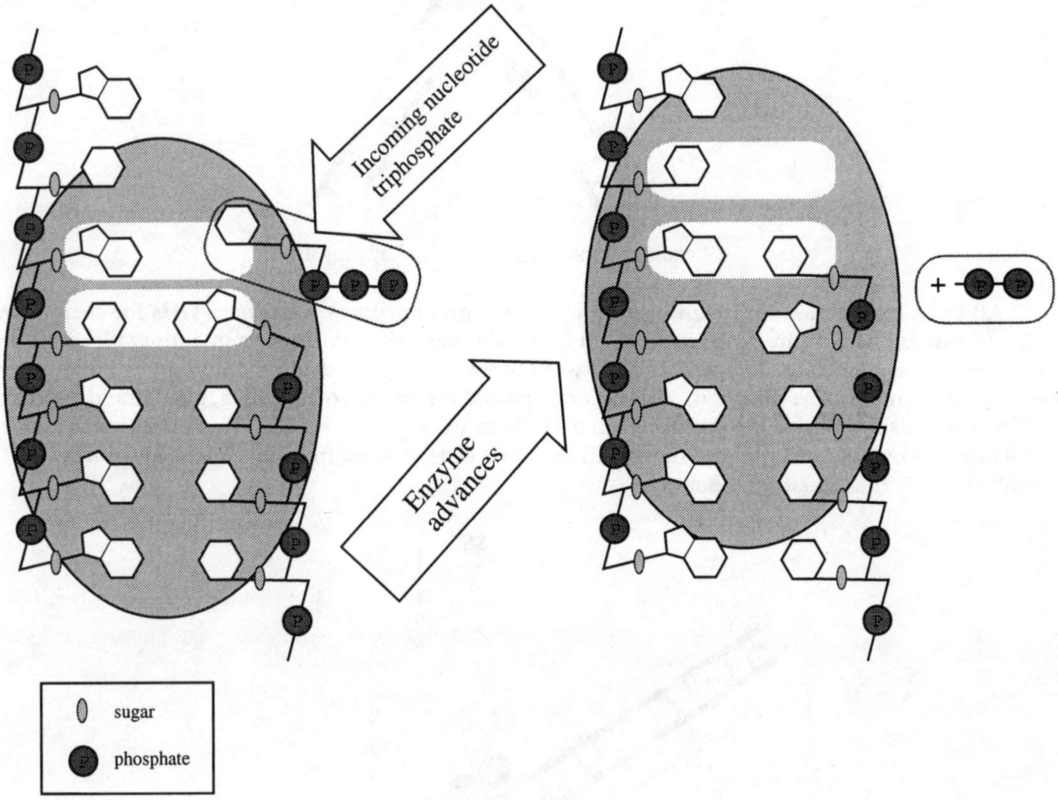

Working from the free 3'OH, DNA polymerase moves into the replication fork. This relatively simple polymerization process is referred to as "leading strand" synthesis. Just ahead of the DNA polymerase is the helicase enzyme, which began unwinding the double stranded DNA at the *ori*. This enzyme works to provide a fresh supply of template for the ever-moving DNA polymerase:

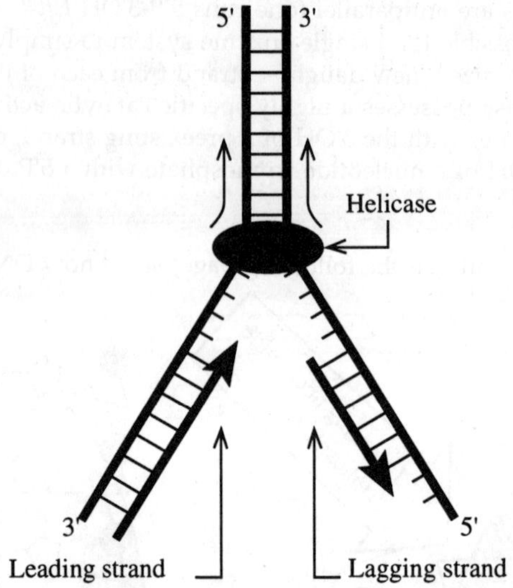

All DNA polymerases are highly specific: They require primers with free 3'OHs for chain growth. This feature of the enzyme is accommodated by synthesizing the two strands in antiparallel directions. As previously mentioned, the antiparallel nature of the two DNA strands does not allow for replication of both strands in the same direction. This is compensated for on the so-called "lagging strand," where a short RNA primer is synthesized in the direction *away* from the replication fork. This primer provides a free 3'OH for use by DNA polymerase III. Pol III then proceeds to work from that 3'OH, away from the replication fork, as illustrated below:

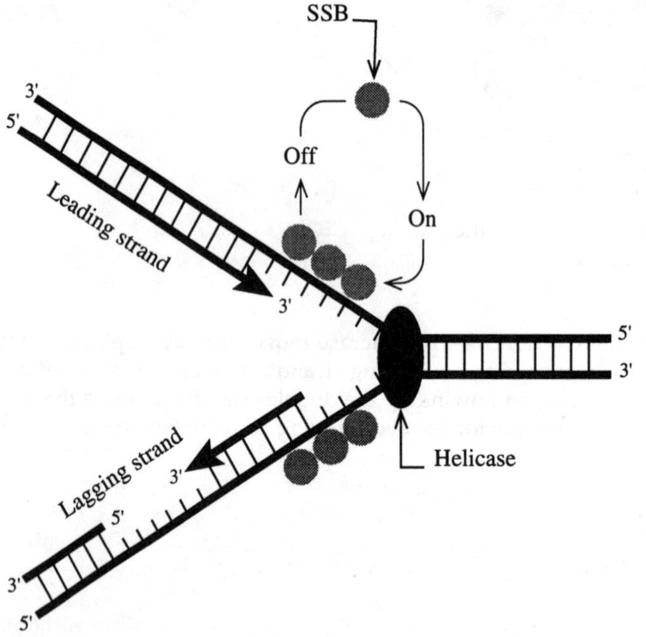

To summarize: one strand is synthesized <u>into</u> the replication fork, while the other is synthesized <u>away</u> from the replication fork.

As helicase continues to unwind the double-stranded helix at the replication fork and the leading strand continues to elongate into the fork, the other strand is being unwound as naked DNA (the first lagging strand having grown *away* from the fork). Quickly, however, another RNA primer is synthesized onto that naked DNA, and Pol III works on the primer's 3'OH end, away from the fork, to synthesize a second segment of the lagging strand.

Eventually, DNA polymerase I (Pol I) edits out the RNA primer by replacing it with bonafide *deoxy*ribonucleotides. A ligase enzyme then joins the 3'OH ends of the individual DNA fragments with the 5'P of the adjacent fragments. This generates a long continuous strand. Lagging-strand synthesis is so called because synthesis actually does lag, in time, several nucleotides behind the rate of polymerization of the leading strand.

Actually, the lagging strand is looped around the Pol III molecule that is busy polymerizing the leading strand. This loop permits a single dimeric Pol-III molecule to polymerize both the leading and lagging strands simultaneously! Figure 7-22 of *EMB* illustrates this loop. Pol III is a very large enzyme. Typically, it consists of several subunits, each of which is present in two copies (a dimer), so it is well equipped to work on both strands at once.

To summarize: Just as we described DNA *replication* as a "semiconservative process," we view DNA *synthesis* as "semidiscontinuous." Why? Because the lagging strand is not synthesized as a continuous strand, but rather as a series of fragments that are eventually linked together. However, the leading strand is synthesized as a continuous strand. Since *one* strand is *not* synthesized as a long continuous strand, the process is referred to as *semi*discontinuous.

When does the rolling circle type of DNA replication come into the picture?

Many viruses contain their entire genome in a single, relatively small circle. For replicating this genome, they use minor modifications of the events just described above. The DNA synthesis process is, therefore, essentially the same for both these small circular viral DNAs and for the much larger, eukaryotic DNAs.

What are those minor modifications?

The first minor modification has to do with the initiation of DNA synthesis. Instead of leading-strand synthesis beginning at an *ori*, where abundant A-T base pairs easily melt (i.e., breathe) and allow replication enzymes to gain access to a single-strand template, a single-strand "nick" is made at a specific site on one of the two strands by an endonuclease enzyme. This nick generates the 3'OH end necessary for DNA polymerase action. In some cases, the bacterial Pol III is used; in others (such as T_4 phage), the polymerase is phage encoded. The DNA polymerase begins to synthesize a leading strand. As synthesis proceeds, the cut strand, in front of the growing leading strand, "peels off" the circle. The unwinding is carried out by a helicase. A "tail" of displaced DNA is generated that dangles off the circle.

The tail also serves as a template for DNA synthesis. But since its polarity (5'P→3'OH) is unsuited for direct action by Pol III (that, you should recall, *always* requires a free 3'OH), synthesis is carried out in the lagging-strand manner. This lagging-strand synthesis is essentially the same as the previous version (described in Figure 7-22), including the necessary primers.

The newly synthesized double-stranded DNA can keep "rolling off" the circle for a long time since there is no obstacle to its continuation. The long strand is then cut by an endonuclease into large pieces and circularized by ligase. The complete circles are then packaged into individual viral protein coats.

To summarize: The viral rolling circle model begins with a single-strand endonuclease cut. Typical leading- and lagging-strand synthesis continues indefinitely. As the long strand of newly synthesized DNA "rolls" off the circle, it is cut into pieces for packaging. This is analogous to an efficient factory assembly line. Many numbers of copies of small circular DNAs can be made quickly. Speed of replication is important for bacterial viruses. They attempt to replicate fast, before the defense mechanism of the bacterial host (e.g., endonucleases) has an opportunity to interrupt the replication process.

What's in a name?

Rolling circle is a neat descriptive name because it allows you to visualize the replication of small circles. One way is to hold the inner, passive, circle fixed in your mind and roll the outer, replicating strand, clockwise around, peeling off the cut strand as you go (strand displacement). Another way is to roll the inner, passive, circle around counterclockwise and keep the position of the dangling tail fixed, except getting longer and longer with time, much like a roll of tape!

Covalent extension is another nice descriptive name for the same process. Since the leading strand is covalently extended, through its phosphodiester backbone, from the original (outer) circle at the nick, this mode of DNA replication is often called the covalent extension method.

Finally, "sigma" replication is another name for this process, because it can easily be drawn to resemble the Greek letter sigma (σ).

Why is prokaryotic DNA usually circular, instead of folded or pleated?

Several advantages accrue DNA when in a circular form rather than a linear form:

1. It can easily supercoil (twist) itself to relieve the physical strain that otherwise might lead to breakage. Also, in a supercoiled state, it can be easily packaged with proteins into a tight, stable format.
2. Large circular DNA can be replicated rapidly by employing several *"origins of replication"*, as will be explained below. The leading strand of one replication fork becomes the lagging strand of the opposite fork in a "replication bubble." There is no danger of Pol III running off the end of a linear molecule and no need for special enzymes to ligate tangled loose ends.

Are there any disadvantages associated with large circular DNA molecules? Yes, one major one, the "topology issue." Special enzymes, topoisomerases, are required to untwist and unlink circular DNAs. It is actually difficult to imagine how tangling is prevented within the close confines of the cell's nucleus. Keep in mind, the typical human cell nucleus contains approximately 6 linear feet of DNA!

Warning!

DNA replication mechanisms will be revisited in Chapter 14 of *EMB*. Leading- and lagging-strand synthesis, rolling-circle replication, and bidirectional replication will all be used to explain plasmid and transposon replication. Please look ahead to Chapter 14. Notice the illustrations in Figures 14-3 and 14-6 of *EMB*.

Key Terms

Semiconservative replication

Topoisomerases

DNA gyrase

DNA Pol I

Nick translation

DNA Pol III

Primer

Holoenzyme

DNA ligase

Discontinuous synthesis

Leading strand

Lagging strand

Pulse-labeling experiment

Pulse-chase experiment

Okazaki fragments

RNA polymerase

Primase

SSB protein

Preprimosome

Ori

De novo initiation

Covalent extension

Bidirectional replication

For More Information . . .

Additional explanations are indeed available:

The entire process of DNA replication is discussed in depth on pp. 285 - 295 in *Molecular Biology of the Gene*, 4th edition, by Watson *et. al.*

It also might be helpful to look at Figures 6-42 and 6-44 in *Molecular Biology of the Cell*, 3rd edition, by Alberts *et. al.*

The Wise Owl Says

Learn from other people's mistakes! Don't waste your time, energy, and money making your own mistakes. The old adage—you learn best from your own mistakes—should be restated: intelligent people learn to avoid pain and problems by observing and learning from other people's mistakes.

Exercise

I. Events at the replication fork

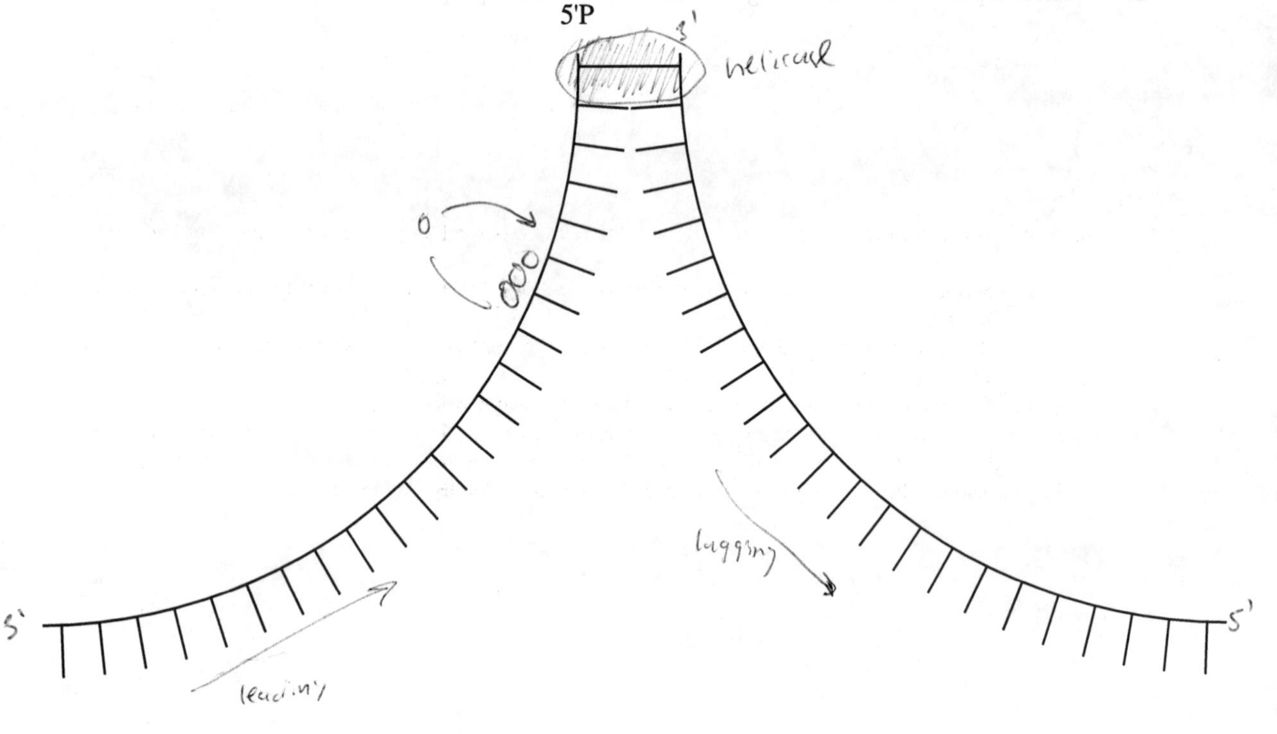

Add the following to the above diagram:

1)	label all strand polarities	6)	primosome
2)	helicase	7)	primase
3)	leading strand	8)	RNA primers
4)	locations of Pol III	9)	locations of Pol I
5)	SSB protein	10)	locations of ligase

P.S.: Please note—getting the 5'P→3'OH polarities correct is the key to proper orientation of all other components.

II. Bidirectional replication

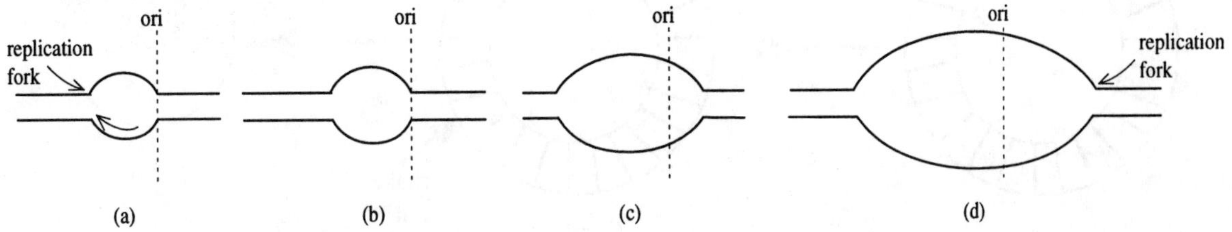

In replication bubble (a), the arrow illustrates leading-strand synthesis that began at the *ori*.

1. Label each of the 4 ends of the double-stranded DNA template with the correct polarities (i.e., 3'OH and 5'P).
2. In (b), illustrate the beginning of lagging-strand synthesis and the extension of the leading strand using arrows.
3. In (c), extend the leading strand and ligate the lagging strands.
4. In (d), illustrate the original leading strand becoming the lagging strand for the right-hand replication fork. Likewise, illustrate the original lagging strand becoming a leading strand.
5. Please double check the 3'OH and 5'P polarities. Are they correct?

III. Rolling-circle replication

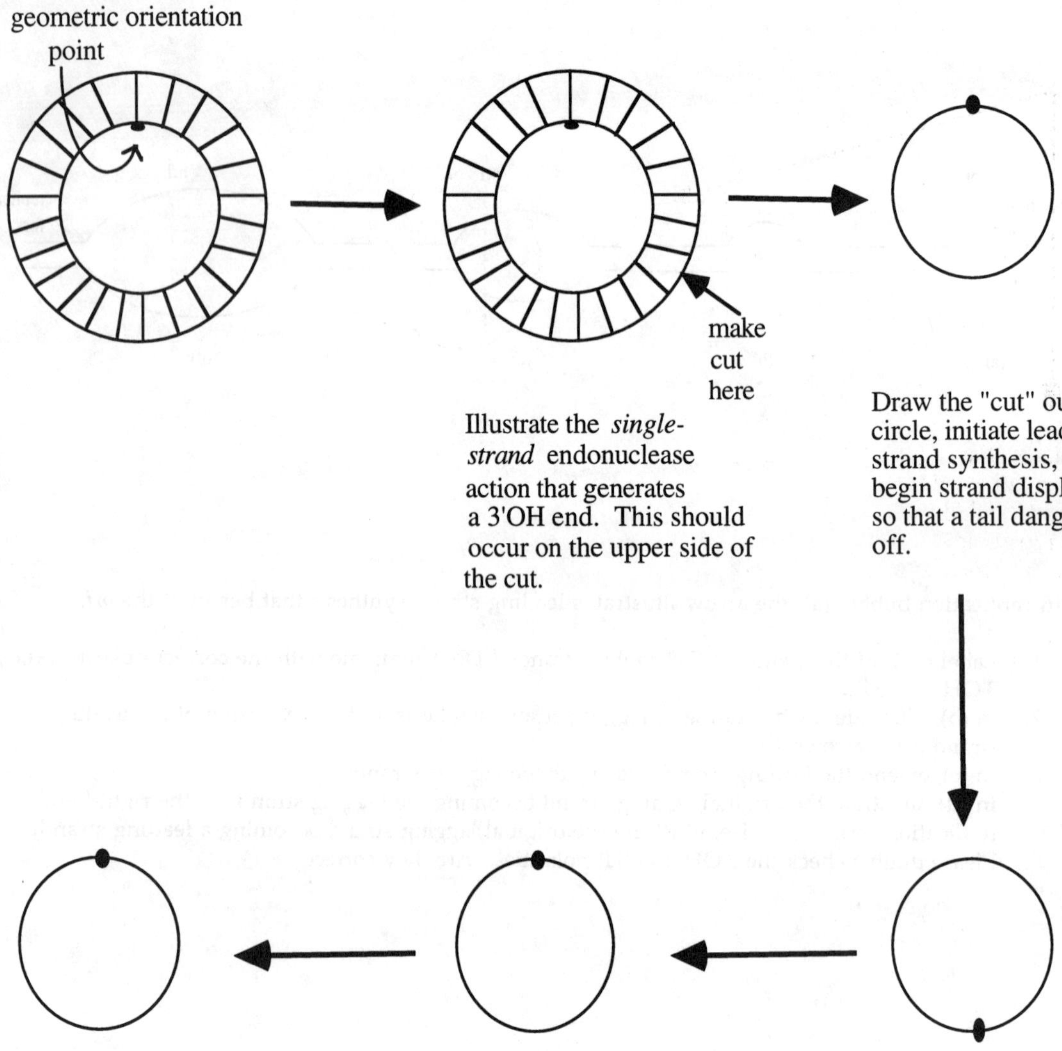

Illustrate the *single-strand* endonuclease action that generates a 3'OH end. This should occur on the upper side of the cut.

Draw the "cut" outer circle, initiate leading-strand synthesis, and begin strand displacement so that a tail dangles off.

1. Extend the leading strand 1/2-turn clockwise around the inner strand.

2. Clearly illustrate the longer tail, indicating displacement.

3. Add the lagging strand replication components to the tail.

1. Extend the leading strand another 1/2-turn clockwise around the inner strand.

2. Illustrate the action of Pol I and ligase on the lagging-strand fragments.

Illustrate the action of the *double-stranded* endonuclease that cuts off the long tail from the *covalently extended* outer circle.

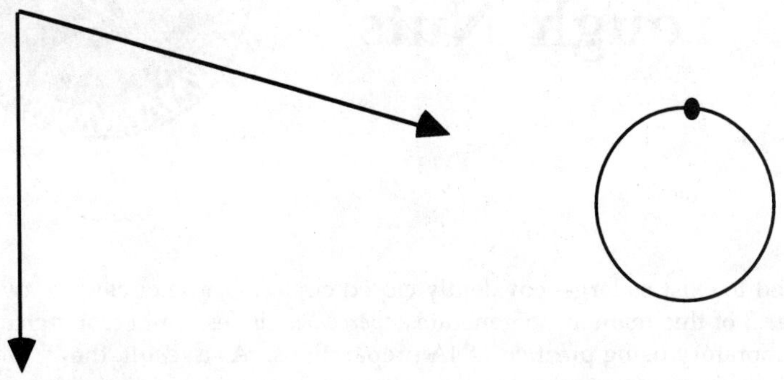
Illustrate how the original leading strand serves as a *template* for additional lagging strand synthesis!
Circularize the newly synthesized double-stranded DNA molecule.

Tough Nuts

I. Prokaryotic DNAs are believed to exist as large, covalently closed circles. These circles are twisted about themselves (see Chapter 3 of this manual) to generate *supercoiled* circles. Supercoiling can be conveniently studied in the laboratory using purified DNA preparations. As a result, the topoisomerase enzymes that introduce coiling and determine the extent to which a circle is supercoiled have been relatively well characterized. But what about supercoiling in the chromatin of eukaryotic cells? Is chromatin DNA highly supercoiled? How large are the supercoiled regions? What role do histone proteins and other chromosomal proteins play in supercoiling? Do topoisomerases function to change the supercoiling in chromatin?

These questions are indeed tough ones to answer in large part because of the complexity of eukaryotic chromatin. In addition to DNA, chromatin contains protein (both histones and other types of "DNA-binding" proteins) as well as RNA. It is very difficult to chemically extract the DNA and isolate it in its "native" configuration. As a result, most knowledge of DNA superstructure is derived from the study of relatively pure prokaryotic DNA preparations, since they are much easier to prepare.

How can we crack this tough nut? At least three possibilities exist:

1. Develop improved extraction methods for eukaryotic DNA, so that it can be isolated intact, in its coiled configuration.
2. Develop methods to assay the various topoisomerase enzyme activities directly on isolated chromosomes.
3. Reconstitute chromatin from isolated components and determine which of those components need to be added to the reconstituted mixture in order to achieve the "native" state of chromatin.

Why would it be difficult to use mutants which have been so useful for various other studies, to analyze *eukaryotic* (e.g., mammalian) DNA supercoiling?

II. Another tough nut lies in the unwinding reaction of DNA at the replication fork. Helicase is the enzyme that does this job. It requires energy, in the form of ATP, and works just ahead of DNA polymerase. The mechanism by which this enzyme manages to unwind and separate one strand from the other in such an orderly fashion is still a mystery.

Three hypotheses currently exist:

1. The "arms" of the replication fork twist around like the blades of a propeller.
2. The arms remain steady while the "soon to be replicated stem" twists around.
3. Phosphodiester bonds are broken, then resealed, with a rotation of the unreplicated region occurring between the breaking and sealing processes.

Fortunately, this process can be studied with prokaryotic organisms, so it might be possible to use mutants to crack this tough nut. What sort of mutants could be used? What will constitute the definitive evidence to prove the correct unwinding hypothesis? Wouldn't it be nice to be able to use physical measurement techniques and actually observe the unwinding? It would probably be necessary to slow down the replication process so that it could be tracked by a physical or chemical measuring device. However, slowing the process down may in itself be a "tough nut." Why not use lower temperatures to slow down enzyme catalyzed reactions? Are there any "downsides" to this approach?

What might be the simplest possible experimental system for studying DNA replication *in vivo*? How about small plasmids that replicate inside bacterial hosts? Chapter 14 in *EMB* discusses them in detail.

Can you think of some other ways to crack these tough nuts?

Study Questions

1. What is the purpose of replication? At what stage in the cell cycle does it occur? What major event does it precede for which it is absolutely necessary?

2. Based on the definition of "semiconservative replication" given in Chapter 7 of *EMB*, is rolling-circle replication semiconservative when it results in three or more copies of the DNA?

3. What is the difference between nick translation and strand displacement?

4. Would a *sudden* loss of **all** gyrase activity in a replicating *E. coli* result in a sudden cessation of replication?

5. Using your knowledge of the synthesis of an RNA primer by the enzyme primase, what is a *major* difference in the "requirements" of primase and Pol I or III?

6. Diagram the various stages of the replication of a linear molecule of DNA from initiation to (near) completion. Do this in detail. In other words, what is shown in Figure 7-10 of *EMB* is *not* sufficient. As always, indicate the polarity of each and every DNA segment.

7. Where are RNA primers needed in rolling-circle replication?

8. Why don't eukaryotes use rolling-circle replication to copy their chromosomes?

9. In addition to Pol I and Pol III, name at least three additional components that must be present for DNA replication to occur.

10. Describe at least three differences in function between Pol I and III.

11. Draw a diagram of DNA replicating bidirectionally. Label the leading strands, *ori*, Okazaki fragments, RNA primers, and the 5' and 3' ends of each and every DNA and RNA strand. Also indicate name, function, and location of at least three of the enzymes involved.

12. Name two enzymes that make primers for DNA synthesis.

13. List the enzymes involved in replication at the replication fork. What other proteins are important?

14. What are some reasons for the complexity of events at the replication fork?

Answers to Chapter 7 Study Questions

1. Provides copies of the genome to daughter cells; occurs during the S (DNA synthesis) phase; precedes cell division.

2. No, since, in addition to the one parental strand, daughter strands serve as templates.

3. In nick translation, the 5'→3' exonuclease activity of Pol I degrades the existing strand one nucleotide at a time as synthesis proceeds at a nick in the DNA. In strand displacement, the existing strand is not cleaved, just displaced.

4. No, because there is a region of single-stranded DNA between gyrase and DNA polymerase. Also, Pol I can unwind the helix as it replicates.

5. Primase requires uracil, whereas Pol I and Pol III require thymine. Also, recall that Pol I and Pol III require a free 3'OH in order to polymerize nucleotides. Primase does not.

6.

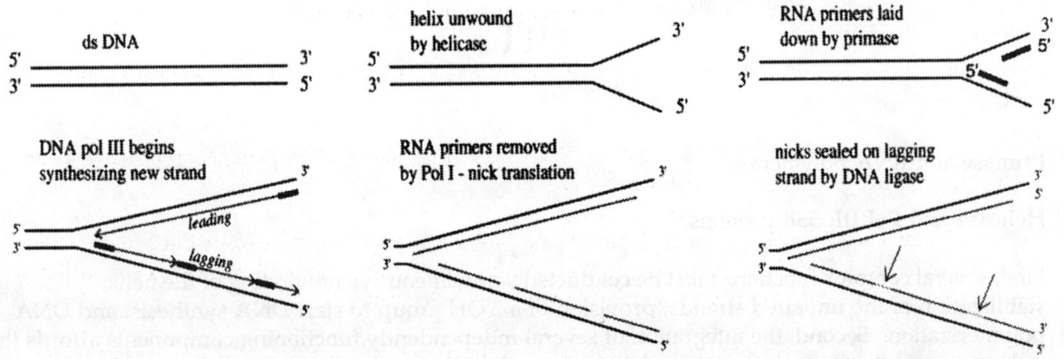

7. For lagging strand synthesis (only).

8. First of all, eukaryotic DNA is linear, not circular. Second, eukaryotic DNA is usually 1,000 to 10,000 times longer than prokaryotic or viral DNA.

9. Deoxynucleotides (triphosphate form); primers (primase); helicase.

10. 1. Pol I has 5'→3' exonuclease activity, Pol III doesn't.
    2. Pol I doesn't require helicase, Pol III does.
    3. Pol I can use ssDNA as a template, Pol III cannot.

11.

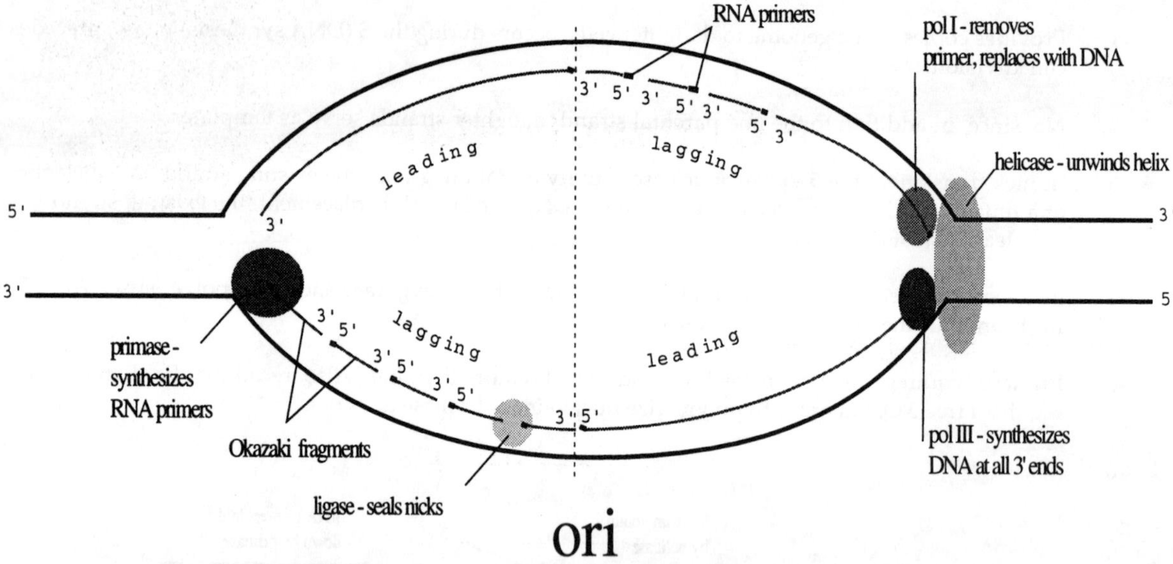

12. Primase and RNA polymerase.

13. Helicase and Pol III; SSB proteins.

14. First, several complex functions must be conducted simultaneously: unwinding of the helix, stabilization of the unpaired strands, provision of a 3'OH group to start DNA synthesis, and DNA polymerization. Second, the integration of several independently functioning components affords the cell more opportunity for regulation of the process than if a single, do-it-all enzyme were used. Third, accurate replication is essential for survival, so precision is necessary, although it certainly adds complexity.

Concept Map

Let's arrange the details of Chapter 7 into a concept diagram:

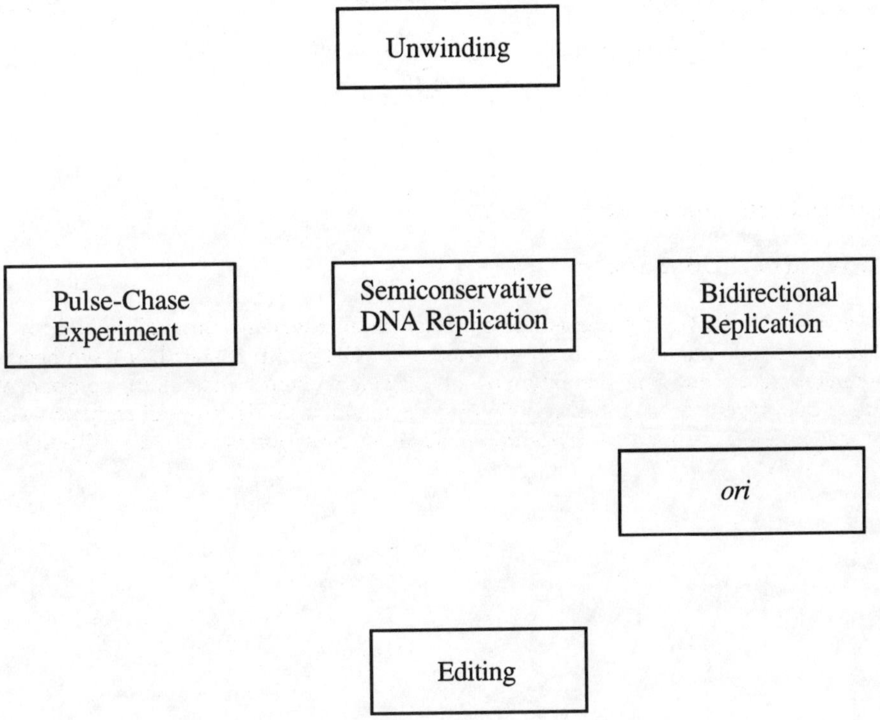

Please be certain that the following components are included in your map:

θ replication
DNA ligase
lagging strand
gyrase
de novo initiation
topoisomerase
leading strand
DNA polymerase
Okazaki fragment
rolling circle

8 Transcription

Here's Help

What are the main steps of transcription in prokaryotes?

1. RNA polymerase (RPol) binds to DNA at the promoter region: the -35 and -10 sequences. At this time, the sigma subunit is attached to the polymerase and is involved in recognizing the nucleotide sequences of the promoter.

2. An open promoter complex forms at the site where RPol is bound. This consists of a 17 base-pair "bubble," or region of unbound DNA. Earlier in this manual, we referred to "breathing" to describe the formation of bubbles in double stranded DNA.

3. The RPol then begins synthesizing the RNA molecule. The first DNA base to be transcribed is usually a T. Therefore, the first RNA base (i.e., at the 5'-P end) will be an A. A ribonucleotide that corresponds to the second DNA base then enters the elongation site of the RPol.

4. The two ribonucleotides are joined together. That is, a covalent, phosphodiester bond is formed between them.

5. The RPol moves 1 base down the template strand, and the next ribonucleotide enters the elongation site. This new ribonucleotide is then joined to the preceding one. By now a 3-base RNA molecule has been synthesized.

6. The RPol continues to move down the DNA, adding ribonucleotides and linking them together.

7. When the nascent (growing) RNA molecule is about 8 bases long, the sigma subunit dissociates from the enzyme. This subunit will, eventually, reattach to another RPol molecule.

8. The core enzyme continues the synthesis of the RNA molecule by adding one ribonucleotide at a time. As the RNA chain grows, only the region associated with the polymerase remains base-paired with the DNA template.

9. Once termination occurs, the polymerase dissociates from the DNA molecule and reassociates with the sigma subunit in preparation for another initiation event. The process of termination is explained below.

What is a promoter?

A promoter is a region comprised of one or more base sequences in a DNA molecule to which RNA polymerase binds. RNA polymerase binds to the promoter at the outset of transcription. The promoter serves to orient the polymerase in the correct direction for transcription.

In prokaryotes, the promoter consists of two distinct sequences just upstream of the site where transcription is to begin. The sequence closest to the start site is called the Pribnow box. This sequence is usually 5'-TATAATG-3', but, regardless, it is always A-T rich. The Pribnow box is located approximately 10 bases upstream of the start site (-10). The other sequence in a prokaryotic promoter is located about 35 bases upstream of the start site and hence is referred to as the -35 sequence.

A eukaryotic promoter usually has three characteristic sequences (plus other ancillary sequences that will be discussed in Chapter 13 of *EMB*). These sequences usually extend approximately 80 bases upstream. One characteristic sequence of the eukaryotic promoter, called the Hogness or TATA box, is similar to the prokaryotic Pribnow box, but differs in that it is located at -17 rather than -10.

Here's a quick review of RNA synthesis termination:

There are two ways by which RNA synthesis is terminated. One way is termed **Rho dependent** and the other is **Rho independent**. In Rho independent termination, a stem-and-loop structure forms in the nascent RNA molecule. The stem of this structure is high in G-C content which facilitates termination by inhibiting melting of this region. A simplified illustration of this structure is included in Figure 8-8 of *EMB*. This stem and loop alone cannot terminate transcription, but must be followed by a series of six to eight uracils. Hence, the stem and loop coupled with the series of uracils serves to halt transcription.

In Rho dependent termination, the Rho protein recognizes and binds to a certain region of the nascent RNA strand. Termination is accomplished through a mechanism that involves the hydrolysis of ATP but is not yet fully understood.

How is eukaryotic mRNA processed?

1. A 7-methylguanosine cap is added to the 5'P-end of the transcript, that is "dangling" from the polymerase, before transcription is complete.

2. Transcription terminates.

3. A poly (A) tail is added to the 3' end of the transcript.

4. The introns are removed.

5. The exons are spliced together.

What are the differences between prokaryotic and eukaryotic mRNA?

One major difference between prokaryotic and eukaryotic mRNA is the lifespan of the molecule. Prokaryotic cells don't live as long as eukaryotic cells. The former live only minutes or hours as compared to the latter, which live days, months, or years. This lifespan discrepancy indicates that most proteins that will be needed during the lifetime of a prokaryotic cell will only be needed in relatively small amounts. On the other hand, it takes much less time to transcribe and translate RNA in a prokaryotic system than it does in a eukaryotic system. The reason for this will be discussed in the next chapter of *EMB*. Thus, it is much more efficient for a prokaryotic cell to make new RNA whenever necessary, rather than "store" and reuse previously synthesized RNA. The mRNA of eukaryotic cells, because of the 5' 7-methylguanosine cap and the 3' poly (A) tail, is more stable and remains in the cell for a longer period of time. Thus, if more of a particular protein is needed, the cell does not have to wait for new RNA to be made.

A second difference between prokaryotic and eukaryotic mRNA is that prokaryotic mRNA is often polycistronic. The regions of the mRNA molecule that code for each particular protein are called cistrons. Different enzymes, involved in steps of the same metabolic process (e.g., degradation of lactose or synthesis of tryptophan) are often encoded by the same mRNA molecule. The polycistronic quality of prokaryotic mRNA expedites the synthesis of all the different enzymes for a particular metabolic process by keeping them under the control of the same regulatory mechanisms.

Yet another difference, posttranscriptional modification, has already been discussed in the previous section. The eukaryotic primary transcript receives a 5' 7-methylguanosine cap and a 3' poly (A) tail. *It also contains introns that are removed by splicing together the exons.* Figures 8-10 and 8-11 of *EMB* illustrate splicing. Those modifications are lacking in prokaryotic transcription.

How about another explanation of Southern blotting (Figure 8-13 of *EMB*)?

Purified double-stranded DNA is first cut into fragments with an endonuclease (DNase). Those fragments are then separated according to their relative sizes in a gel electrophoresis system (e.g., see Figure 2-3 in Chapter 2 of *EMB*). The separated fragments appear as "bands" on the gel, and are transferred to a piece of nitrocellulose paper, using a weight (to compress the gel) and "buffer flow" to move the bands out of the gel and into the filter paper.

Once *blotted* or *transferred* onto the nitrocellulose paper, a "filter hybridization" procedure of the type described in Figure 3-12 of *EMB* can be carried out.

Why is it called "Southern" transfer?

The scientist who first developed the technique is Dr. Edward Southern!

Key Terms

Holoenzyme

Core enzyme

Promoter

Consensus sequence

-35 sequence

Open-promoter complex

Cistron

Leader

Intrinsic terminators

Rho protein

Ribosome

Probe

Primary transcript

Posttranscriptional modification

5' 7-methylguanosine cap

3' poly (A) tail

Introns

Exons

Heterogeneous nuclear RNA

RNP

Domain

Hybridization

Clone

Southern blot

For More Information . . .

Do you need additional explanations on these topics?

Please consult pp. 224-227 5 of *Molecular Biology of the Cell* by Alberts *et. al.* (3rd Edition) for another description of the process of transcription.

For a good explanation of the hybridization and Southern blotting techniques discussed in this chapter, consult pp. 608-609 of Watson *et. al.*, *Molecular Biology of the Gene*, 4th edition.

The Wise Owl Says

A "brilliant" idea that isn't explained properly is about as useful as a supercharged race car without one of its wheels.

As molecular biology research becomes more sophisticated and better able to unravel the complex metabolic processes of living cells, communication skills become more valuable.

Molecular biologists' research agendas have been increasingly comprised of complex phenomena that require accurate, precise, and clear explanations. Enhancing your verbal and written communication skills will pay rich dividends as your career in molecular biology develops!

Exercise

I. Use the formulae on the following page to illustrate the chemical synthesis of a short RNA molecule. Be certain that the 5'-P and 3'-OH ends are clearly indicated. The key to this exercise is to illustrate phosphodiester bond formation. Recall, phosphodiester bond formation involves a hydrophilic "attack" on the 5'-triphosphate of the incoming nucleotide by the 3'-OH that was last added to the growing RNA. Formation of the phosphodiester bond results in the elimination of a pyrophosphate from the incoming nucleotide triphosphate.

Draw the formula of a short RNA segment that contains approximately 6 nucleotides.

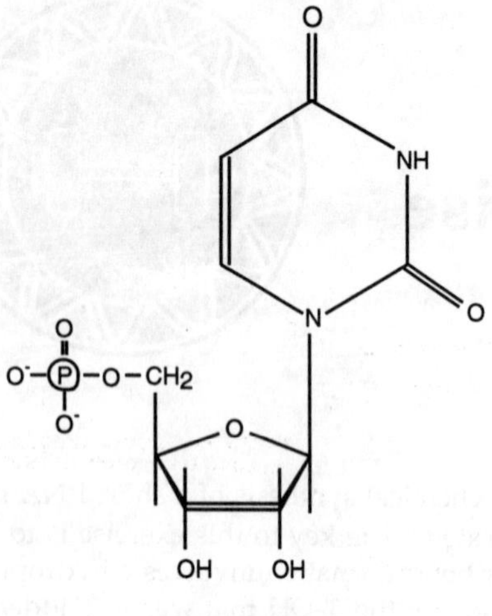

Uridine- 5'-phosphate

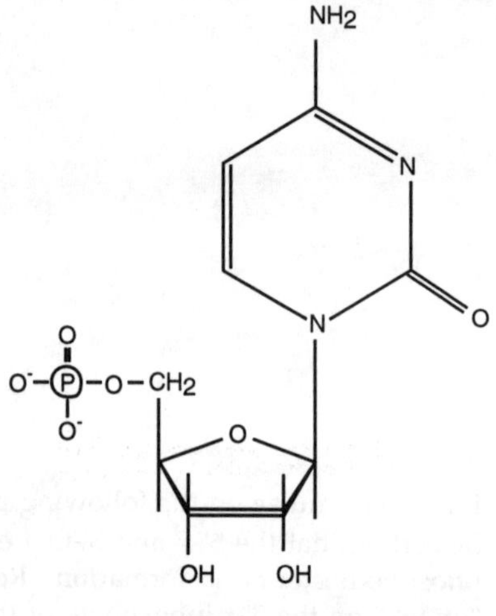

Cytidine-5'-phosphate

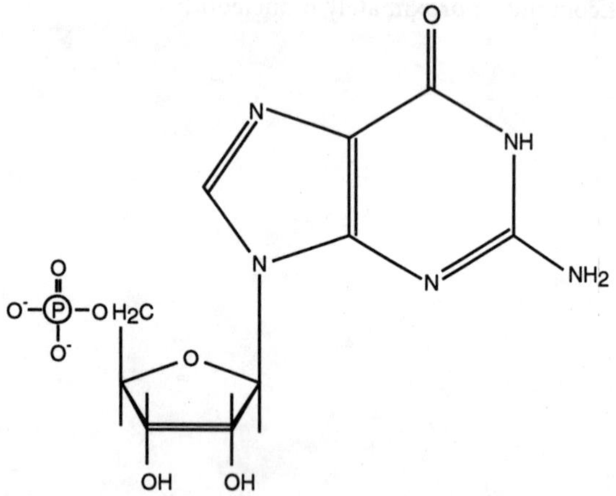

Guanosine- 5'-phosphate

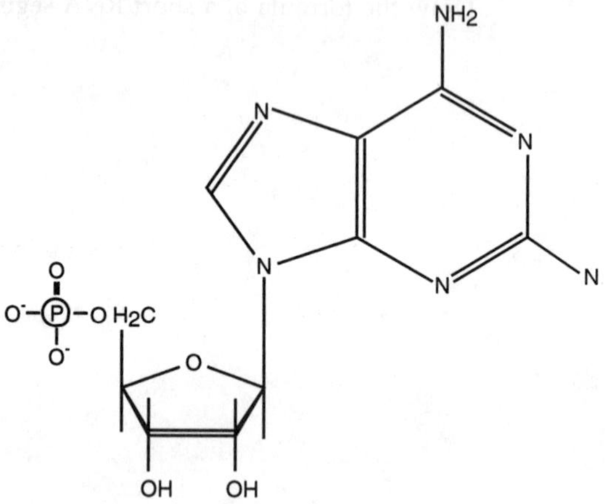

Adeninosine- 5'-phosphate

II. Add labels to the following transcription complex:

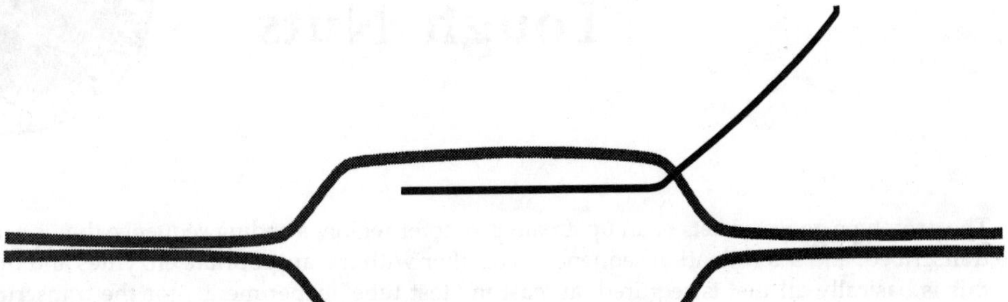

Labels to be added:

*5'-P and 3'-OH of DNA template

*5'-PPP of growing RNA strand

*3'-OH of growing RNA strand

*hydrogen bonds between complementary bases

*an arrow indicating the direction of RNA polymerase movement

III. Illustrate the promoter region of a gene by drawing a 3-D diagram of the interaction between the polymerase and the DNA. Be certain the following key points are included in your drawing:

1. The double helix must make a complete turn every 10 nucleotides (see Figure 3-1 in Chapter 3 of *EMB*).
2. The enzyme must make contact with the DNA at two sites: approximately the -10 and -35 nucleotide regions.
3. The enzyme's contact with DNA must be limited to only one side of the double helix.

Tough Nuts

I. The prototype gene consists of an upstream promoter region, a coding sequence that is actually transcribed, and a termination sequence. Together with the appropriate enzymes and nucleotides, this is basically all that is required, at least in "test tube" experiments, for the transcription of mRNA to occur. However, *in a natural setting things appear to be much more complicated.* For example, additional proteins are thought to be involved in some organisms. Mammalian RNA polymerase is a case in point. It may consist of 8-10 subunits. Additional sequences that are outside of the DNA coding region may also play a role in altering the frequency with which transcription of the coding sequences occurs. Such sequences often reside far upstream from the coding sequence. In other instances, they reside in the intervening sequences (introns) of a gene. Yet others are probably located downstream. Some of these sequences appear to enhance the frequency of transcription initiation and, hence, are called "enhancers" (you will encounter these later, in Chapter 13 of *EMB*). These enhancer sequences probably represent recognition sites for specific proteins that bind to the nucleotide sequence and thereby facilitate initiation of transcription. Other sequences act to diminish the frequency of transcription initiation and are called "silencers." However, the action of these sequences is, in general, poorly understood.

Establishing the complete inventory of upstream, and possibly downstream, nucleotide sequences that affect gene expression (i.e., transcription initiation) is a tough nut. At least two approaches to cracking this tough nut are being attempted:

1. Exhaustive nucleotide sequencing, of regions far removed from the gene of interest, are being carried out in order to recognize putative protein recognition sequences. Once recognized, both genes that do and those that do not contain the sequences are analyzed in "test tube" experiments to determine which sequence speeds up, or slows down, transcription. It is, however, a tough job. How far away from the coding sequence should one search?

2. DNA folding is being investigated in order to establish whether nucleotide sequences at a distance can fold back and exist in close proximity with the gene's upstream promoter. Such folding would provide a basis for speculating how a DNA binding protein that complexes with a "remote" sequence acts to facilitate the binding of RNA polymerase to its promoter sequence. However, these studies are also difficult. Is DNA, when it is being transcribed in a living cell, free of all contact with histones? If not, how much freedom of movement and flexibility for folding does DNA have in its natural environment? Tough stuff!

II. Many of the details of transcription have been revealed during the course of studies with relatively "simple" organisms. That is, invertebrates, such as nematodes and fruit flies, have proved to be especially amenable for studies based on the molecular genetics of transcription. Indeed, most of the nucleotide sequences that comprise the sites where "regulatory" proteins (see Figure 5-12 in Chapter 5 of *EMB*) are known to bind have been discovered in these organisms.

Will identical, or even similar, sequences be present in human genomes? Pharmaceutical companies hope to be able to design drugs that can bind to those sequences. The rationale behind this is to either "up-regulate" or "down-regulate" the activity of disease causing genes.

To what extent will the background information provided by nematodes and fruit flies be useful for human molecular genetics studies? Scientists argue both "for" and "against." A tough nut, indeed.

Can you propose a way to crack these tough nuts?

Study Questions

1. What are the subunits that comprise RNA polymerase? What are their functions?

2. What specific DNA sequences are involved in prokaryotic transcription?

3. Are transcription initiation sequences needed between the cistrons of a polycistronic mRNA?

4. RNA polymerase, unlike DNA polymerase, has no proofreading function. Why is this function important for DNA polymerase, but not RNA polymerase?

5. Given the mRNA sequence 5'-GCCAAUGCUCGGAAUAACGCCU-3', why would the UAA sequence near the 3' end not act as a stop codon?

6. What are some of the eukaryotic posttranscriptional modifications of the primary transcripts of tRNA, rRNA, and mRNA?

7. Compare RNA polymerase with DNA polymerase III. What enzymatic activities does one have that the other does not?

8. Design an experiment to see if RNA polymerase molecules, after having finished transcription, rebind to the same sigma subunit that dissociated earlier in transcription.

9. Propose a method to measure the "lifespan" of newly synthesized RNA molecules.

10. Why don't RNA polymerases need the SSB proteins that are required for DNA replication?

11. If you hybridized a strand of DNA to the strand of RNA transcribed from it, would the complex be parallel or antiparallel?

12. In prokaryotes, how does the primary transcript differ from the mature mRNA molecule? What about in eukaryotes?

Answers to Chapter 8 Study Questions

1. 2 identical α subunits, 1β subunit, and 1β' subunit come together to form the core enzyme. The σ (sigma) subunit joins the core enzyme to form the holoenzyme. The σ subunit functions primarily in orienting the holoenzyme.

2. Pribnow box, -35 sequence, start sequence, termination sequence.

3. No

4. DNA must have high fidelity if genetic information is to be retained over many generations. RNA, on the other hand, lasts only minutes or hours before it is degraded, making fidelity less important.

5. Because it is not in the same "reading frame" as the AUG start codon. This is explained in great detail in the next chapter (9) of *EMB*.

6. tRNA—endonuclease cuts at various positions on the 5' and 3' ends of the molecule, various base modifications within.

   rRNA—excision of a single continuous sequence.

   mRNA—MeG capping, addition of poly (A) tail, and the cutting and splicing of exons.

7. Both DNA pol III and RNA pol catalyze the 5'→ 3' addition of nucleotides using an anti-parallel template stand. RNA pol lacks the 3'→ 5' editing function of DNA pol III. RNA pol can initiate chain growth without a primer, which also contrasts with DNA pol III.

8. Allow transcription to occur in a test tube with a surplus of sigma factor. After time has elapsed, to allow all RNA pol to become associated with sigma factor, add radioactively labeled sigma factor. After some time, separate the DNA-RNA pol complex from the solution. A decrease in the level of radioactivity in the solution would indicate that some of the newly added sigma factor was associated with RNA pol, refuting the hypothesis that the same sigma factor binds to the same RNA pol continuously.

9. The "lifespan" of an mRNA molecule could be determined by measuring the length of time that it produces functional protein in a cell. Therefore, measure the enzymatic activity of the protein product after adding an RNA pol inhibitor to the cell. The amount of time required for half of the original activity to be lost is the half-life of the mRNA (this assumes no loss of protein activity over time).

10. RNA polymerase is much larger than DNA polymerase. It catalyzes the melting of the DNA double helix as well as keeps the strands from reannealing at the site of RNA synthesis. DNA polymerases have neither of these abilities.

11. Antiparallel

12. In prokaryotes, the primary transcript is the mature mRNA molecule; no posttranscriptional processes occur. In eukaryotes, the primary transcript is up to ten times as long as the mature mRNA molecule. This is due to the excision of introns and splicing of exons. Also, the mature form has a MeG cap on the 5' end and a poly (A) tail on the 3' end.

Concept Map

A Concept Map will undoubtedly be helpful in understanding this chapter. Shall we prepare one?

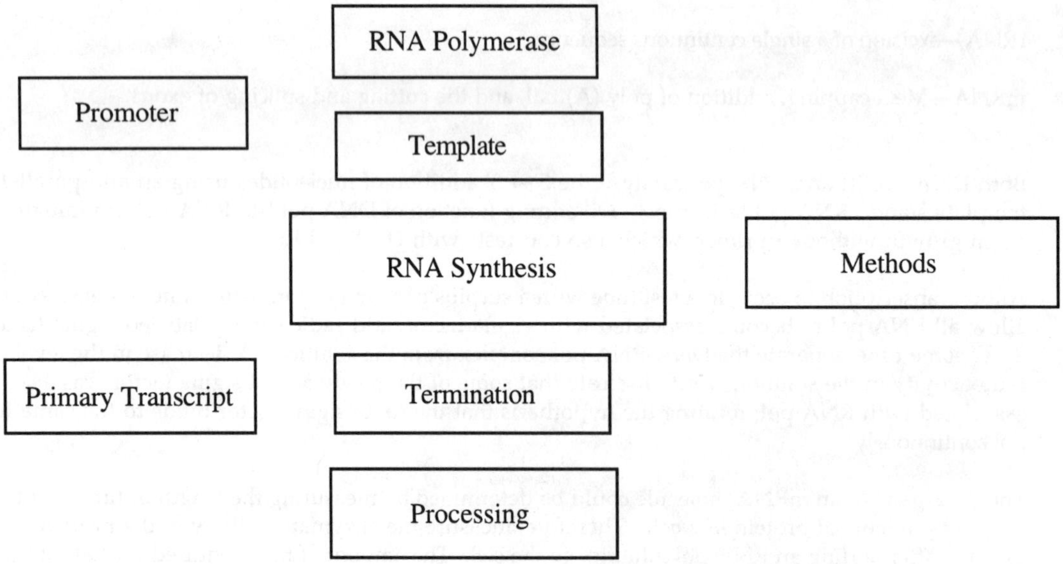

Please include the following components in your concept map:

- tRNA
- holoenzyme
- mRNA
- introns
- cap/tail
- splicing
- probe
- core enzyme
- terminators
- sigma subunit
- ribosomal RNA
- exons
- Southern transfer

9 Translation

Here's Help

Why is the direction of transcription and translation so important?

RNA polymerase moves along the coding strand of DNA beginning at the 5' end. Thus, the mRNA molecule grows by addition to the 3' end (the 5' end is synthesized at the beginning of the transcription event). Translation is also initiated at the 5' end, and the ribosome moves along the RNA molecule in the 3' direction. Thus, the growing end of the polypeptide is the C terminus. (The first amino acid comprised the N terminus.) This polarity makes the coupling of transcription and translation possible in prokaryotic cells. Figure 9-1 of *EMB* illustrates these directionalities. Since the ribosome begins translating the 5' end of the transcript, it can start before transcription is completed. This permits many ribosomes to translate an mRNA molecule at the same time, forming a polysome. Hence, the rate of protein synthesis in prokaryotic cells can be greatly accelerated., for even before an mRNA is fully-synthesized, its translation can begin.

Here's a brief review of "wobble base-pairing":

"Wobble base-pairing" is a descriptive term used to explain codon-anticodon pairing that deviates from the strict base-pairing we have previously observed throughout *EMB*. The pairing between the anticodon loop and the DNA's triplet of bases is somewhat nontraditional. The first two codon positions base-pair in the expected fashion. The third codon position, however, allows some slight mismatching between the codon and anticodon. This "wobble" exists because it is not necessary for all 3 bases in the anticodon loop of the tRNA molecule to form the traditional A-U and G-C pairs. For example, the anticodon of a tRNA that contains an inosine (I) in the first position can base-pair with an A, U, or C (see Table 9-2 of *EMB*) in the third position of the codon. (Remember that only in the first position of the anticodon are adenosines changed to inosine.) Likewise, a G of the anticodon's first position can pair with either a U or a C. Therefore, the third base of the codon is less influential in defining the triplet code than are the first two bases. A quick review of Table 9-1 should convince you of this. Note how the amino acids that are represented by multiple codons usually contain identical first and second bases, but different third bases!

Why are so many prokaryotic genes transcribed as polycistronic messenger RNAs?

Polycistronic mRNAs, as illustrate in Figure 9-6 of *EMB*, , code for more than one protein. A single long mRNA often is transcribed from several adjacent genes. This means that the protein products of the adjacent genes can be synthesized in a coordinate fashion, and all of the proteins can appear in the living cell at approximately the same time. Usually, polycistronic mRNAs code for proteins that carry out related functions. For example, several enzymes that work together in the synthesis of specific amino acids are encoded in a single polycistronic mRNA molecule.

In Chapter 1 of *EMB*, the logic of molecular biology was explained. Perhaps polycistronic mRNAs represent an example of the "efficiency argument " that is often invoked to explain "why processes work the way they do." Could the presence of "overlapping genes" (Figure 9-7 of *EMB*) be explained in the same way?

What role do spacer sequences play in the regulation of protein synthesis?

Spacers (e.g., Figure 9-6 of *EMB*) are noncoding regions between cistrons on a polycistronic mRNA molecule. Remember that cistrons of the same mRNA molecule often code for enzymes involved in the same metabolic pathway. Let's assume that we know of a pathway that involves three different enzymes: A, B, and C, encoded by cistrons *a*, *b*, and *c* of the same RNA molecule. Also assume that the amount of activity required from each of the enzymes fits a 3:2:1 ratio (meaning that in the living cell A is used three times as much as C and B twice as much as C). If all three cistrons are translated the same number of times, then there will be a wasteful excess of B and C. For example, if the cell needs 3000 molecules of A, it would have to make 3000 of each A, B, and C. However, because of the 3:2:1 ratio, only 2000 of B and 1000 of C are needed. This is where spacers come into play.

It was originally assumed that all ribosomes translating cistron *a* also translated *b* and *c* (located downstream from *a*). This, however, was eventually proven to be incorrect. In reality, there are spacer regions between the cistrons in a polycistronic mRNA. Ribosomes tend to "lose their grip" on the RNA molecule when the untranslated spacer regions move through the ribosome. The longer the spacer region, the greater the chance that a ribosome will detach. This is how the spacers regulate the ratio of products. For this to be successful, the cistron encoding the most needed product must be the first cistron on the molecule (from the 5' end). The product needed in the second greatest amount must be the second cistron, and so on.

In our example, the spacer between cistrons *a* and *b* would have to be of such length that if 3000 ribosomes were to complete translation of cistron *a*, then approximately 1000 ribosomes would have to drop off the molecule before reaching cistron *b* due to the length of that spacer region. To create the desired 3:2:1 ratio of products, of the remaining 2000 ribosomes that translated cistron *b*, approximately 1000 more would have to drop off when the spacer region between cistrons *b* and *c* move past. This would leave only 1000 of the initial 3000 ribosomes to actually translate cistron *c*. This is an example of how spacers can help achieve a certain ratio of products. Remember, however, this only works on proteins that are encoded by cistrons located on the same mRNA molecule.

Are "introns" and "spacers" related in any way?

No, not at all. Introns represent "intervening" sequences in a gene (DNA) of a eukaryotic cell which are transcribed along with coding sequences into a primary transcript. They are then cut out of the primary transcript before translation of that mRNA begins.

Spacers are nucleotide sequences which are of course present in a gene (DNA) of a prokaryotic cell, but also included in the mRNA which is translated.

A simple way to make that distinction is to associate introns with eukaryotic DNA and spacers with prokaryotic DNA

How does initiation of translation differ in prokaryotes and eukaryotes?

In prokaryotic mRNA, the Shine-Dalgarno sequence (Figure 9-12 of *EMB*) designates the nearest downstream AUG codon as the start codon. The 16S rRNA of the 30S ribosomal subunit has a complementary sequence to the Shine-Dalgarno sequence. The binding of these two complementary sequences initiates prokaryotic translation.

In eukaryotes, the AUG codon that acts as the start codon is simply the one closest to the 5' end of the molecule. The ribosome binds to the 5' end and then slides along until this first AUG codon is reached.

What steps are involved in the translation of a prokaryotic mRNA?

1. The 30S subunit, a molecule of GTP, an fMet-$tRNA^{fmet}$ molecule, IF-2 (IF = initiation factor), IF-3, IF-1, and the mRNA molecule associate to form the initiation complex. IF-3 is released once this complex has been formed.

2. The 50S subunit joins the complex, GTP is hydrolyzed to GDP, and IF-1 and IF-2 are released.

3. Now, the ribosome exists with the fMet-tRNA in the P site. The A site, already located "over" the next codon, is still empty.

4. A charged tRNA molecule, with an anticodon that matches the codon in the A site, enters the A site. This step is mediated by EF-Tu (EF = elongation factor).

5. Two aminoacyl tRNAs are now complexed with the ribosome. The carboxyl terminus of the amino acid in the P site (fMet) forms a bond with the amino terminus of the amino acid in the A site. The bond between the amino acid in the P site (fMet) and its tRNA molecule is now broken. These events are catalyzed by the enzyme peptidyl transferase.

Note: The state of affairs is this: A dipeptide (two amino acids) is bound to the tRNA in the A site and the tRNA, without an amino acid, in the P site.

6. Three reactions now occur simultaneously: The uncharged tRNA in the P site exits from the ribosome, the peptidyl-tRNA in the A site moves to the P site (still remaining base paired to its complementary codon in the mRNA, and the ribosome moves three bases down the mRNA. The codon that was in the A site has been moved to the P site, and the next codon has been moved into the A site. This translocation step requires the participation of elongation factor EF-G, and is "fueled" by the hydrolysis of GTP.

   Note: The next steps are simply repeats of steps 4 through 6.

7. A charged tRNA molecule, with an anticodon that matches the new codon in the A site, enters the A site.

8. The amino acids attached to the tRNA in the P site are transferred to the amino acid attached to the tRNA in the A site. Thus, the amino acid chain is increased by one link.

   Note: Now an uncharged tRNA exists in the P site, and a tripeptide is bound to the tRNA in the A site.

9. The three-step translocation occurs again (see Step 6).

10. This cycle continues until a stop codon moves into the A site. No tRNA has an anticodon that corresponds to a stop codon, so translation can, therefore, go no further. The stop codon is recognized by a protein release factor (RF), which causes the polypeptide chain to be released from the tRNA in the P site.

What prevents the mischarging of tRNA?

Most amino acids are easy to distinguish from one another based on the different chemical properties of their side chains. For example, amino acids such as lysine, glutamic acid, and phenylalanine, which have highly distinctive side chains, very infrequently misacylate tRNA molecules. However, the chance of misacylation of the tRNA is greatly increased by amino acids with side chains that have very similar chemical properties, such as valine and isoleucine, or glycine and alanine. In order to avoid wasting valuable molecules, the tRNA synthetases for these molecules have evolved ways to detect and correct this problem.

The isoleucine tRNA synthetase provides us with a good example of this built-in editing function. After attaching an isoleucine molecule to $tRNA^{Ile}$, it moves the newly charged tRNA molecule to the "proofreading sector" of the enzyme. This proofreading niche has a "pocket" that is shaped like the side chain of valine, since valine may be mistakenly attached to $tRNA^{Ile}$. The isoleucine molecule correctly attached to the tRNA won't fit into this pocket and is, therefore, released from the enzyme. When the enzyme does make a mistake and charges $tRNA^{Ile}$ with valine, the valine side chain will fit into the proofreading pocket, and this charged tRNA will not be released from the enzyme as before. Instead, the tRNA synthetase is activated to hydrolyze off the valine molecule from the $tRNA^{Ile}$ in order to correct the mistake. Neat, eh?

How about an *informal* interpretation of Figure 9-16 (coupled prokaryotic transcription and translation)?

The diagram in Figure 9-16b of EMB interprets the electron micrograph in Figure 9-16a The synthesis of mRNA 1 is just beginning, whereas the synthesis of mRNA 4 is almost complete. Both are still, however, physically linked to DNA (the slanted vertical line) by the enzyme RNA polymerase.

Ribosomes hop onto the mRNA (at its 5' end [where the synthesis began]) and begin translating proteins. The proteins are so tiny and threadlike that they do not show up in the electron micrograph in (a), and therefore are not shown dangling off the ribosomes (as they are in Figure 9-15 of *EMB*).

The synthesis of protein could be illustrated by combining Figures 9-15 and 9-16: flip Figure 9-16 over so that the 5' end of the mRNA is on the right (in Figure 9-15 it is situated on the left side), and lay it down on mRNA 2 in Figure 9-16. That will show the synthesis of proteins!

Key Terms

Translation

Transfer RNA

Anticodon

Aminoacyl tRNA synthetase

Acylated

Uncharged tRNA

Mischarged tRNA

Wobble hypothesis

70S ribosome

Initiation factors

Shine-Dalgarno sequence

30S preinitiation complex

70S initiation complex

A and P sites

Peptidyl transferase

EF-Tu

EF-G

Release factors

Spacers

Polysomes

Coupled transcription-translation

For More Information . . .

Do you need additional explanations on these topics?

A more detailed explanation of the wobble concept can be found in Watson *et. al.*, *Molecular Biology of the Gene*, 4th edition, on pp. 438-440. The illustration on p. 439 displays some examples of wobble pairing.

Diagrams that illustrate ribosome structure and the A site-P site story are included on pp. 231-234 of Alberts *et. al.*, *Molecular Biology of the Cell*, 3rd edition.

The Wise Owl Says

...if I tend to view the simple diagram and the attractive generalization with a critical eye, it is because a very varied experience has made it clear to me that, with rare exceptions, information comes slowly and is accompanied, more often than not, by a great deal of misinformation.

HENRY HARRIS
Oxford
Michaelmas 1967

P.S.: The details of protein synthesis are very well understood. A massive effort by molecular biologists, begun in the 1960s, led to the rapid accumulation of knowledge on this subject. Research was aided by the fact that protein synthesis could be carried out in a test tube. This meant that precise experiments could be performed in which various components were added or subtracted from the reaction mixture. Also, the use of inhibitors of protein synthesis, especially antibiotics such as puromycin, streptomycin, and chloramphenicol, provided valuable insights.

Nevertheless, if we were to leaf through the scientific journals of the 1960s and 1970s, we would probably be surprised at how much misinformation was accumulated along the way to discovering the details of protein synthesis!

Exercise

I. Wobble

Let's increase our understanding of the wobble concept. Please draw out a DNA base sequence containing the six different leucine codons:

5'P ____ 3'OH

Now, add the anticodon loops of the tRNAs and clearly illustrate the expected base-pairings of the codon-anticodon.

Refer to Table 9-2 of *EMB* and simplify the above illustration by replacing, wherever possible, the base in first position of each anticodon with the "all purpose base-I."

Q. When normal base-pairing was employed, how many different tRNAs were required to service the six different leucine codons?
A. ____________

Q. How many different tRNAs were required when wobble base-pairing was employed?
A. ____________

Some types of living cells appear to economize on their tRNA inventories by employing wobble base-pairing. Neat, eh!

II. tRNA charging

Amino acids are covalently attached to tRNA by an ester bond of the amino acid's carboxyl group to the hydroxyl group of the ribose of the last base (adenine) at the -CCA terminus of tRNA. In order to fully understand this bonding, let's draw out the chemical structure of aminoacylated-tRNA molecules. We will begin with a simple case: Glycine.

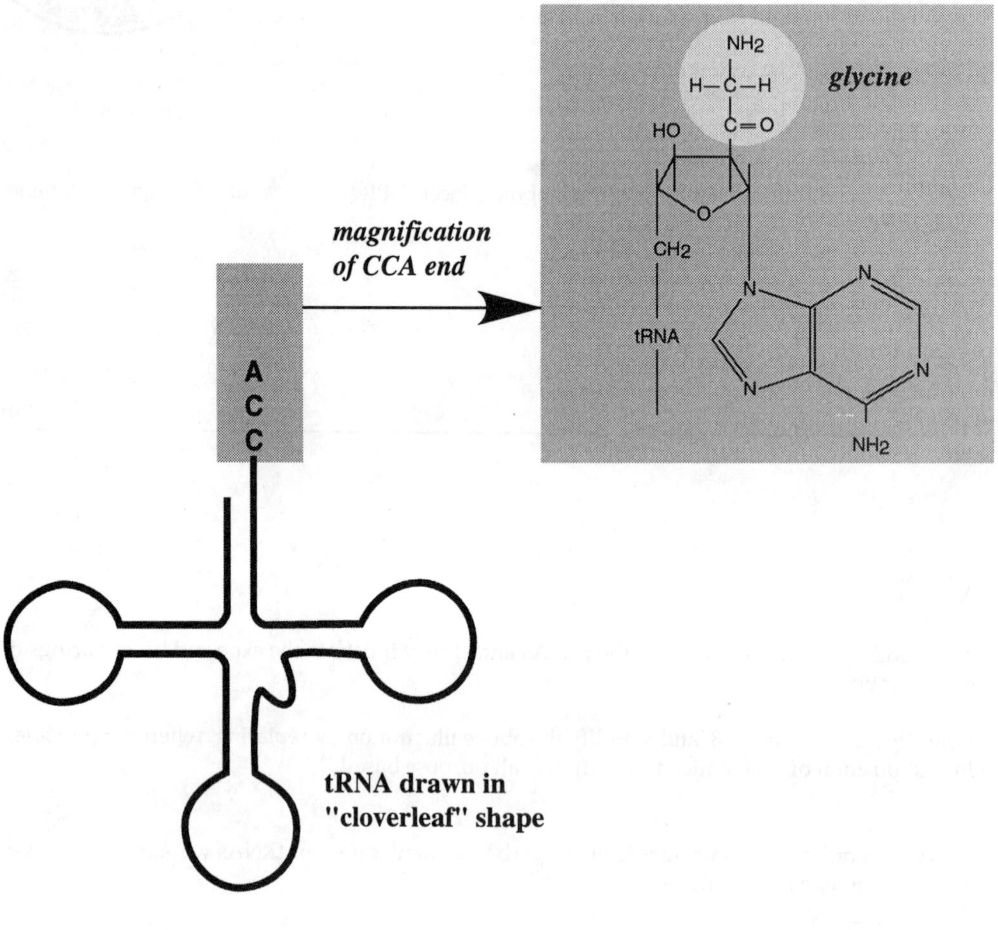

Now let's practice. Draw a somewhat expanded aminoacyl-tRNA formula (-CCA end). In addition to the adenine nucleotide, include the two cytosine nucleotides and glutamic acid bonded to the terminal nucleotide. For another example, replace glutamic acid with phenylalanine.

III. Peptide Bond Formation

By drawing in the required components, including the charged tRNA molecules you illustrated previously, show how a tripeptide is formed. Focus on the action at the P and A sites, and show movement of the ribosome along the mRNA.

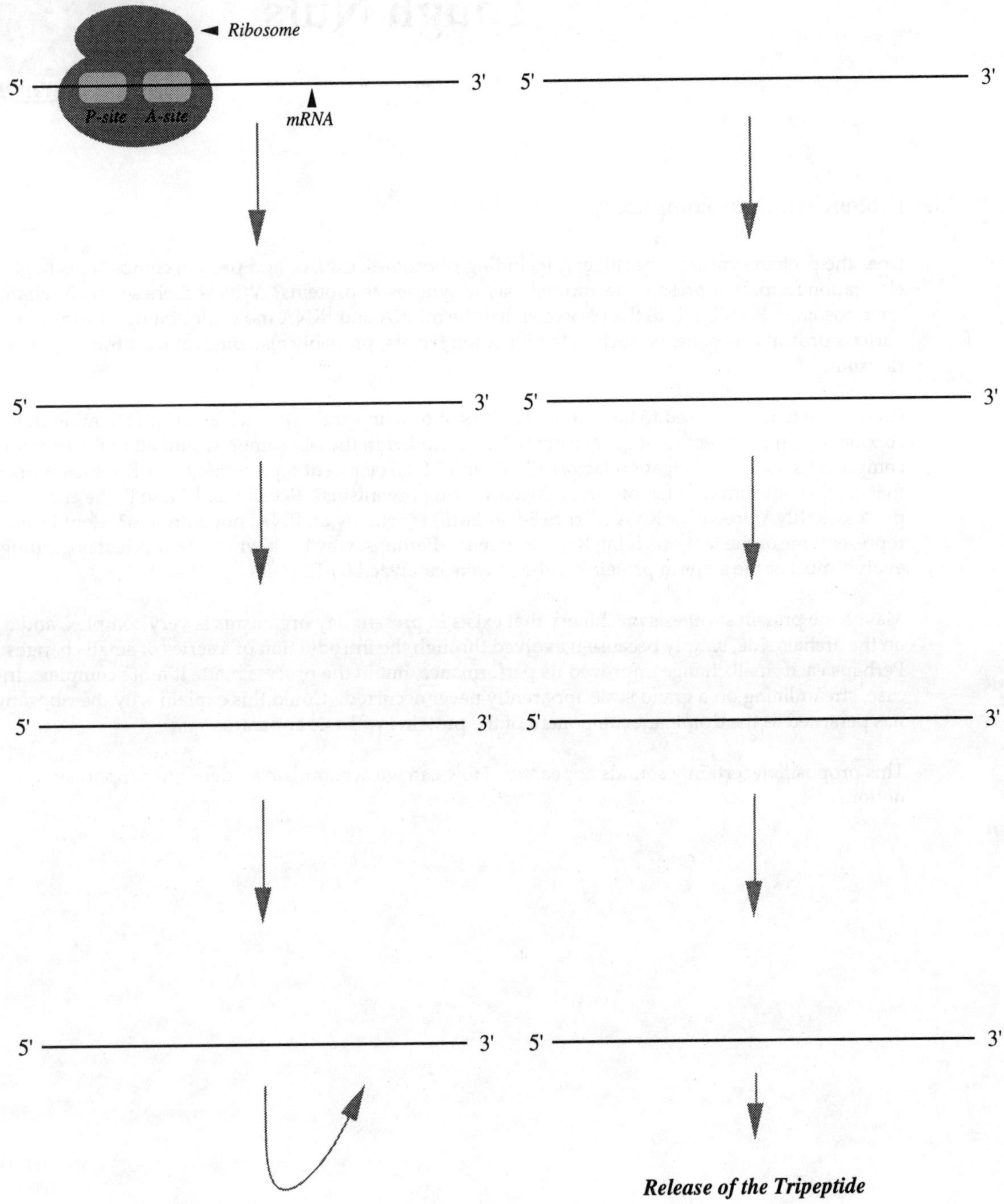

Tough Nuts

I. Protein synthesis complexity

Does the protein synthesis machinery, including ribosomes, tRNAs, and protein components (e.g., elongation factors), represent the optimal way to synthesize proteins? Why is there so much reliance on the ribosome? Recall, it is to the ribosome that the mRNA and tRNA molecules bind. In addition, various protein components, such as the initiation factors, probably also bind, at least transiently, to the ribosome.

Could it be that there used to be even more steps of protein synthesis, such as the information now contained in the nucleotide sequence of mRNA, included in the ribosome? Could all of the protein components, like the elongation factors EF-Tu and EF-G. be recent replacements for RNA components that carried out similar functions in prehistoric living organisms? Recall that RNase P, the enzyme that processes tRNA precursor RNA (Figure 8-9 in *EMB*) is actually an RNA, not a protein? Could this represent one of the last surviving RNA enzymes? Perhaps, way back when life was just beginning to evolve, most of the steps in protein synthesis were catalyzed by RNAs!

Maybe the protein synthesis machinery that exists in present day organisms is very complex, and a bit on the archaic side, simply because it evolved through the introduction of a series of small changes. Perhaps each small change improved its performance, but in the process made it more complex. In this case, streamlining on a grand scale apparently never occurred. Could this explain why the ribosome has persisted as the dominant component of the protein synthesis apparatus?

This proposition certainly sounds appealing. How can we accumulate evidence to support such a notion?

II. Ribosome structure

Molecular biologists remain puzzled over the size and sequence complexity of rRNA molecules included in fully functional ribosomes. Let's take a quick glance at the prokaryotic 16S rRNA molecule. This molecule is then combined with a collection of proteins to form the functional 30S subunit of the fully mature 70S ribosome.

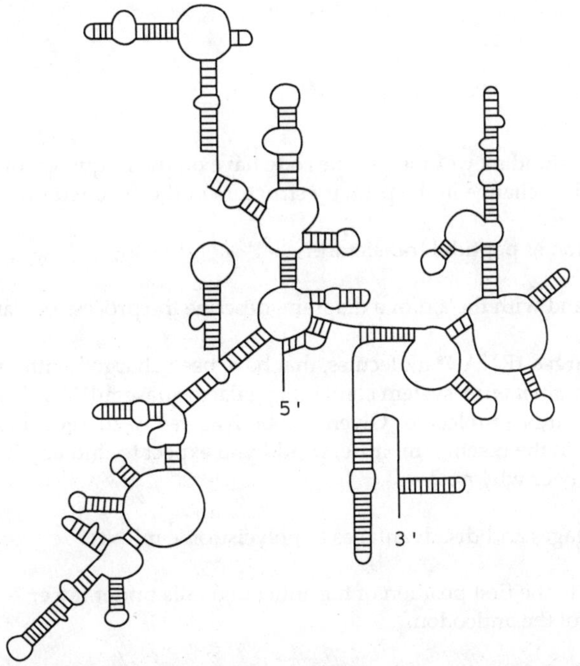

Each fold, loop, and circle is just a representation of the most likely shape a 16S rRNA molecule would fold itself into, based on its nucleotide sequence. Since rRNA sequence has been highly conserved throughout evolution, it is safe to say that these shapes have a functional significance. However, the relationship of the folds, loops, and circles to ribosome function is still poorly understood.

One idea concerning this relationship is that the shapes serve as binding sites, or anchoring places, for the various proteins that interact with rRNA. Therefore, it is thought that the shapes contribute both to the folding process itself and to the long term physical stability of the fully formed 30S ribosomal subunit.

Another idea is that some of these shapes (e.g., circles) serve as binding sites for other RNA molecules that bind to ribosomes. The unpaired bases in the circles can easily be imagined to base-pair with single stranded regions of other RNA molecules.

Yet one more idea is that ribosomes function in more precise ways than are presently understood. Perhaps each fold and loop serves as a site where a discrete, minor step in protein synthesis occurs.

How to crack this tough nut? Why not synthesize novel rRNA molecules that are missing specific loops or circles and examine their functional capabilities? Perhaps a certain portion of the rRNA molecule can be correlated with a specific function.

A tough nut indeed!

Study Questions

1. What effect does the redundancy of the genetic code have on the frequency of a single-base change mutation that results in a change in the primary structure of the encoded protein?

2. What is the key function of peptidyl transferase?

3. In a concise manner, and with the aid of a diagram, describe the process of translation in prokaryotes.

4. Suppose some mischarged $tRNA^{Lys}$ molecules, that have been charged with radioactive serine instead of lysine, were used in an *in vitro* system containing galactosidase mRNA. There are several lysine codons present in the mRNA molecule. Given that both correctly-charged and mischarged $tRNA^{Lys}$ molecules are present in the reaction mixture, would you expect to find any radioactive serine in the resulting protein? Why or why not?

5. Describe some advantages and disadvantages to polycistronic mRNA?

6. Explain why the base in the first position of the anticodon falls under fewer "constraints" than the bases in the other positions of the anticodon.

7. What amino acids are bound to the tRNA molecules that recognize stop codons in a wild-type cell?

8. How do prokaryotes and eukaryotes differ in the selection of a start codon?

9. Which amino acid is present at the amino terminus of an unprocessed polypeptide chain in prokaryotes and eukaryotes?

10. Why is inosine never found in the anticodon position that corresponds to the first base of the codon?

11. How do the ribosomes of eukaryotes and prokaryotes differ?

12. What would be the effect on a prokaryotic system if transcription were suddenly to occur in the 3' to 5' direction?

13. What are some reasons that translation cannot be coupled to transcription in eukaryotes?

14. Which , do you think, would occur more frequently: misacylation of $tRNA^{Ala}$ with glycine, or misacylation of $tRNA^{Phe}$ with tyrosine?

Answers to Chapter 9 Study Questions

1. For those amino acids represented by multiple codons (e.g., thr, leu, or ala), a single-base change does not necessarily change the amino acid since the mutated codon might specify the same amino acid. This is not the case for amino acids represented by single codons (e.g., met or try). In these instances, a single-base change will generate a coding sequence that specifies a different amino acid.

2. It forms a peptide bond between the two amino acids that occupy the P and A sites on a ribosome.

3. Your diagram should include the following features: formation of an initiation complex; ribosomes; GTP; initiation factors; fMet-tRNA^{fmet}; P site; A site; mRNA; charged tRNA; elongation factors; peptidyl transferase.

4. Yes, but probably not much, since only some of the tRNAs are mischarged. Therefore, only some of the "lysine slots" would be filled with serine.

5. The two main advantages are: (1) economy of nucleotide sequences, since very little "spacer" DNA is employed; (2) efficiency of regulation, since the expression of all the genes represented in the polycistronic mRNA can be easily regulated in a coordinated fashion. Two disadvantages include: (1) highly sophisticated regulation mechanisms that involve expressing each gene in *individual amounts* are precluded; (2) evolutionary change, which often includes duplication of *individual* genes, is constrained.

6. The first base in the tRNA anticodon pairs with the third base of the codon. Because the anticodon loop is large and flexible, as long as the first two bases in the codon are properly paired with the second and third bases of the anticodon (tRNA), the pairing is sufficiently stable for peptide bond formation.

7. None. The lack of an amino acid leads to termination of protein synthesis.

8. In prokaryotic mRNA a special nucleotide sequence, the Shine-Dalgarno sequence, is located close to the AUG. This combination serves as the start codon. In eukaryotes, formation of the initiation complex involves the binding of the initiator tRNA to the P site of the 40S ribosomal subunit prior to mRNA binding . The small ribosomal subunit binds to the 5' cap end of the mRNA, then slides down the mRNA until the first AUG codon is reached.

9. Since the amino terminus is the initiation site for protein, methionine, the amino acid that corresponds to the AUG start codon, represents the first amino acid in the polypeptide chain. In prokaryotes, the initiating methionine is formylated, whereas in eukaryotic proteins, it is not. However, in both cases, the methionine is usually removed after the protein is synthesized.

10. Inosine provides flexibility in base-pairing. However, if in the anticodon position for the first position of the codon, this flexibility would lead to mistakes in tRNA base matching with the mRNA.

11. Eukaryotic ribosomes are basically similar to prokaryotic ribosomes, except they are somewhat larger. They contain some additional proteins, as well as an additional (small) RNA molecule.

12. Because protein synthesis is tightly coupled with mRNA synthesis, that is ribosomes attach to the mRNA before the mRNA's synthesis is complete, chaos would result. The sequence, now read "backwards," would most likely result in the formation of nonsensical proteins.

13. First, the nuclear membrane serves as a physical boundary line; ribosomes are not allowed in the nucleus, so protein synthesis could not occur as the mRNA is synthesized. Second, the eukaryotic mRNA is synthesized as a primary transcript, which is then extensively processed (e.g., downsized).

14. Since glycine is structurally more similar to alanine than tyrosine is to phenylalanine, the former mistake would probably occur more often.

Concept Map

Concept Map time! Let's be certain we understand how the details of translation relate to one another:

mRNA

tRNAs

Ribosomes

Genetic Code (Information)

Polypeptide Synthesis

Triplets

Antibiotics

Polycistronic mRNA

Include the following components in your concept diagram with connecting lines and arrows:

- anticodon
- peptidyl transferase
- release factors
- overlapping genes
- wobble hypothesis
- elongation
- polysome
- aminoacyl tRNA synthetases
- initiation factors

Practical Applications

It has been demonstrated that a large assortment of medically prescribed antibiotics operate by inhibiting a specific step in protein synthesis. Obviously, some antibiotics work exclusively in prokaryotes, while others work only in eukaryotes due to the differences in protein synthesis between the two. Those antibiotics that interrupt only prokaryotic protein synthesis have been most extensively exploited. These include streptomycin, an inhibitor of the initiation step, erythromycin, an inhibitor of the peptidyl transferase step, and puromycin, an inhibitor of the peptide elongation step.

These antibiotics are very important tools for molecular biologists because their action leads to a "traffic jam" during protein synthesis. The intermediates that "back-up" ahead of the traffic jam can then be characterized. Several of the steps in protein synthesis were elucidated in this way due to the existence of such a wide range of antibiotics that inhibit protein synthesis.

The resulting information concerning the steps of protein synthesis has, in turn, been used by molecular biologists to improve the effectiveness of naturally occurring antibiotics. That is, by taking advantage of the increased knowledge about the details of protein synthesis, improved versions of antibiotics have been synthesized.

In addition, understanding the mechanisms of action of commonly used antibiotics has aided molecular biologists in their attempts to learn how bacteria become resistant to antibiotics. Antibiotic resistance is an important medical problem. As resistant strains of bacteria appear, novel antibiotics are designed to combat them. Then in turn these "novel" antibiotics become obsolete as yet new resistance mechanisms are displayed by the pathogenic bacteria.

10

Mutational Change and DNA Repair

Here's Help

What is meant by a "conditional" mutation (Chapter 10 in *EMB*)?

In the case of a conditional mutation the mutant phenotype (e.g., abnormal appearance of specialized structures) is exhibited only under certain conditions. Just as the name implies, the mutant phenotype can only be detected when the organism is exposed to appropriate circumstances (i.e., appropriate "conditions").

Most often, the "condition" refers to temperature. At the *restrictive* temperature, the product of a mutant gene, presumably a protein, probably undergoes a conformational change that renders it functionally inactive. At the *permissive* temperature, the gene product functions normally. By being able to adjust the temperature, thereby controlling the expression of the mutant phenotype, the experimenter can manipulate the mutant organism, allowing it to grow at the permissive temperature, or to become abnormal at the restrictive temperature. By shifting the temperature back and forth, it is often possible to gain insight into the normal function of a gene in the life cycle of a multicellular organism. It is also possible to propagate the mutant organism at its permissive temperature and then switch to its restrictive temperature in order to better observe its mutant phenotype. This feature is very advantageous for the experimenter, as you might imagine.

What is the purpose of the mismatch repair system, given that the polymerases already have an editing function?

The only function of the mismatch repair system is to remove any incorrect bases added by DNA polymerases I and III that were not subsequently removed by the editing functions (3' –> 5' exonuclease) of those enzymes. Recall from Chapter 7, Pol I and III add nucleotides to the growing strand. When nucleotides are added that don't base-pair with the corresponding base on the template (parent) strand, the polymerase can proceed no farther. Therefore, the editing function must be used to remove the unpaired base or DNA synthesis stops for good. However, sometimes mistakes do go uncorrected, and an incorrect base may be inserted into the growing molecule by Pol III. This is where the mismatch repair system comes into play. It corrects mistakes that were missed by the editing function of Pol III.

Need a further explanation for the role of methylation in mismatch repair?

If you were to extract the DNA from a prokaryotic cell that was just about ready to replicate its genome, but had not yet begun, you would notice that many of the cytosine and adenine residues have methyl groups (-CH_3). These methyl groups are not present on newly synthesized DNA strands. This provides the enzymes of the mismatch repair system a way of distinguishing old, parental, or template, DNA, from new, daughter DNA upon encountering a double-stranded DNA molecule. The strand that has just been synthesized will contain fewer methyl groups than the strand that served as the template during replication.

Why is it necessary for the mismatch repair system to distinguish between the daughter strand and the template strand? Well, when repairing mutations it is assumed that the correct sequence is that of the parental strand. When the enzymes of the mismatch repair system discover incorrectly paired bases, they need to know which base to remove. The key to the enzymes' distinguishing between the strands is the degree to which each strand is methylated. The strand with more methylation is recognized as the template and left intact, while the undermethylated strand is acted upon.

What are mutational hot spots?

DNA is normally methylated at certain positions. This occurs for many reasons; some were just discussed in the preceding section. Other reasons will be presented later, in Chapter 13. In prokaryotes, DNA is methylated on the A in the sequence GATC. In eukaryotes, however, DNA is methylated on cytosine residues in the sequence 5'-CG-3'. In Chapter 10 of *EMB* it is also mentioned that cytosine occasionally deaminates. When this occurs in unmethylated cytosine, uracil is formed. This is easily detected as being incorrect and is thus repaired by editing enzymes. However, when methylated cytosine is deaminated, thymine is formed. Since thymine is a normal component of DNA, incorrect thymines are not detected, and, therefore, the mutation is not corrected. For reasons which will be discussed later, in Chapter 13, the CpG sequences tend to be clustered in certain areas of the DNA. Since cytosines in these particular clusters are often methylated, the chances are high for mutations to occur in these regions. These regions have therefore been called "mutational hot spots."

How do base analogs act as mutagens?

Base analogs are nucleoside triphosphates that closely resemble the natural bases found in DNA. Due to this similarity (e.g., Figure 10-5 in *EMB*), they can easily be incorporated into DNA during routine replication processes. Once incorporated into DNA, they may undergo a *tautomeric shift* (i.e., isomerize in such a way that they becomes capable of an alternative form of base pairing), and fail to base-pair accurately with the original DNA strand. This leads to errors in DNA replication and transcription because of the lack of perfect base pairing . Figure 10-6 of *EMB* illustrates a consequence of incorporation of the base-analogue 5-bromouracil (5-BU) into DNA. In its normal form, 5-BU acts as a base analogue of T and base-pairs with A. After undergoing a tautomeric shift, however, 5-BU resembles C and is capable of base pairing with G. Although both normal bases and base analogues are capable of undergoing tautomeric shifts (see Figures 10-1 and 10-5), this process occurs much more frequently in the base analogues. This is the reason that base analogues are such effective mutagens!

What is a "mutator" gene?

Mutations in some genes (four types are listed in Chapter 10 of *EMB*) lead to an overall increase in the frequency of mutation. How could this happen? The first example described is perhaps the easiest to understand. Suppose a mutation occurred in the gene that encodes the DNA polymerase enzyme. Let's assume for the sake of example that this mutation results in diminished exonuclease function in that enzyme. When editing newly synthesized DNA, the enzyme, because of its defective exonuclease activity, will be unable to repair many errors. Hence, the newly synthesized DNA will contain an increased number of unrepaired mistakes, or mutations.

The DNA polymerase gene is, therefore, called a "mutator" gene because when it itself is defective, the overall frequency of the occurrence of mutations becomes elevated.

What are the different types of reversion and suppression?

To better understand reversion, let's begin by reviewing the molecule that is being reverted. In most cases, a mutated enzyme is reverted and is hence restored to normal function. Here's a typical situation: Enzyme X doesn't function because a point mutation has occurred in the DNA that encodes enzyme X. (Remember, a point mutation is a single-base change.) Suppose the cell suddenly starts making normal enzyme X. What could have happened? A reversion or suppression of the mutant base sequence most likely occurred: *Another mutation occurred in the DNA encoding enzyme X that restored the formerly mutant product to full, or nearly full, activity.* The event which restored function could have occurred *within the same gene as the original mutation* (**intragenic**), or it might have occurred within a *second, different gene* (**intergenic**) . Let's consider each of these possibilities. (1) The second mutation could have occurred in the same base as the first mutation, changing that base back to the base that was originally present in the wild-type gene. Alternatively, since the genetic code is redundant, and most amino acids are specified by more than one codon, a second mutation may have occurred within the same base triplet of the DNA as the original mutation, changing the mutant codon to a codon that specified the same amino acid as the original, wild-type codon. This type of reversion is called *same-site reversion* because the reverting mutation occurred at the same site as the original mutation. A 100% recovery of function is usually observed in organisms that have experienced this type of reversion; (2) The second mutation could have occurred in a different base in the same general enzyme encoding region of the DNA as the base that experienced the first mutation. This second mutation somehow manages to compensate for the first mutation. This type of mutation is called a *second-site* or *suppressor mutation,* and represents an example of **intra**genic suppression Since second-site reversion does not correct the mistake but instead just "covers it up" or more accurately stated, "compensates for it," organisms that have experienced this type of reversion usually do not function as well as the wild-type organism.

Now let's contrast these **intragenic** events with other **intergenic** events that could bring about suppression of the mutant phenotype. Remember, enzymes can be monomers, made up of only one protein subunit and so encoded by only one gene, or they can be multimers, made up of more than one subunit and so may be encoded by more than one gene. When dealing with a multimeric protein, it is possible for a second-site suppressor mutation to occur that is not only located at a different site than the original mutation, but that is on a completely different gene. This is referred to as *intergenic suppression.* Intergenic suppression is a special type of second-site "reversion" that may occur when the enzyme is a multimer encoded by more than one gene. The mutation occurring in the second gene is able to compensate for the original mutation that occurred in the original gene. On the other hand, *any* reverting mutation, either same-site or second-site in either a monomer or dimer, that occurs on the *same* gene as the original mutation is referred to as an *intragenic reversion.*

How does a nonsense suppressor mutation work?

Nonsense suppressor mutants can be explained in terms of nonsense suppressor tRNAs. Chapter 10 of *EMB* provides such an explanation. The original mutation is thought to have caused a codon change in the mRNA that codes for a protein, converting a codon that specifies an amino acid to a stop or "nonsense" codon. Translation of the mRNA containing such a nonsense mutation comes to a halt, and a truncated, usually-nonfunctional mutant protein is produced. Why? Because there are no *normal* tRNA molecules that have anti-codons capable of base-pairing with stop codons. Thus, no amino acid can be inserted into a growing polypeptide chain when a stop codon is encountered. Now, suppose that a *second, intergenic mutation* occurred in a tRNA-coding gene, specifically in the region of that gene that coded for the anti-codon loop of the tRNA molecule. If this tRNA gene was mutated in such a way that its *mutant anticodon loop* was complementary to the particular stop codon present in the mutant protein-coding mRNA, the cell would now have available a tRNA molecule capable of recognizing this stop codon and inserting an amino acid into the growing polypeptide chain at the point at which this stop codon appeared. Thus, the second mutation in the gene that codes for the altered tRNA compensates for the original mutation. The original mutation still exists, but is serviced by a mutant form of tRNA. This mutant tRNA "suppresses" the original nonsense mutation, hence the name "nonsense suppressor" tRNA.

How about a thumbnail sketch of the Ames test, described in detail in Chapter 10 of *EMB* ?

The Ames test is a very sensitive technique that allows us to determine whether a substance is a potential carcinogen by looking at the ability of the substance to cause mutations in bacteria. In this case, the "mutation" is a *reversion* (=reverse mutation) from *His-* phenotype to wild-type.

P.S.: The sensitivity is extraordinarily high because of several factors which were built into the design of the test, including (1) use of both base substitution and frameshift *His*$^-$ mutants (which exhibit unusually low rates of spontaneous reversion); (2) agar plating on medium lacking histidine (which allows us to recognize even a few revertants in a large population of bacteria); (3) use of leaky test bacteria (so most potential carcinogens can penetrate the bacterial membrane); and (4) the addition of mammalian liver extract to the culture medium (to "activate" potential carcinogens which might ordinarily *become* carcinogenic as a result of "processing" by enzymes present in the mammalian liver).

What is excision repair?

Thymine dimers cause the shape of the DNA double helix in which they occur to be distorted. This is illustrated in Figures 10-11 and 10-14 of *EMB*. That distortion causes problems in the replication and transcription of the double helix. Excision repair reduces those problems by removing and replacing the segment of DNA containing the thymine dimers. In the excision repair pathway, an endonuclease makes two single-strand cuts in the backbone of the DNA molecule, one on either side of the thymine dimer. However, these cuts are not necessarily made directly adjacent to the thymine dimer. Thus, the strand with the dimer would look something like this:

3'___5' 3'_____TT________5' 3'_____5', where the polarity of the termini and the ends of the gap are shown, and TT represents the two thymines of the dimer. Remember, we are dealing with only one of the two strands of the DNA molecule. The next step is carried out mainly by DNA polymerase I. In Chapter 7 of *EMB* we learned that this enzyme has the ability to carry out strand displacement. Beginning at the free 3'OH group adjacent to the dimer, Pol I synthesizes a new section of DNA while displacing the old, dimer-containing segment. The nick is then ligated and the strand is back to normal! Figure 10-14 of *EMB* illustrates this process.

Why is the double-stranded DNA helix so well-suited for repair?

It can be argued that the nucleotide bases in DNA are well-suited for repair. For example, occasionally cytosine spontaneously deaminates to uracil (see Chapter 6 in *EMB*). When this happens, the proofreading enzymes that scan the DNA are able to easily recognize the U since its presence in DNA is clearly unusual. In fact, it is generally believed that RNA evolved before DNA because of the "inferiority" of U as a base. If DNA contained U, the editing mechanisms that recognize U would not be able to function, because a deamination of a C→U would not "stand out" among all the other U's. So, presumably DNA evolved a synthesis system and structure that employs thymine (T) as the fourth base instead of uracil (U).

One of the other four nucleic acid bases, G, can be explained in a similar fashion. Adenine (A) deaminates to hypoxanthine (H), a base with features similar to guanine (G). Hypoxanthine can base-pair with cytosine (C). So why does DNA use G rather than H? If H were the natural base-pair mate for C, the DNA repair enzymes would get confused when spontaneous deamination of A occurred to generate an H. How would they distinguish the new H from the other naturally occurring H's? By employing G instead of H, the problem of recognizing spontaneous A deamination products is circumvented. All H's are simply removed!

Thus, it appears that the four nucleotide bases (A, G, C, and T) are indeed well-suited for repair when viewed in the context of the living cell that contains RNA as well as DNA.

When does "repair" really not mean true repair?

Please don't get confused by what is called "SOS repair" (Figure 10-17 in *EMB*). It is really not a mechanism of repair at all! When a polymerase comes upon a thymine dimer during replication, it is physically blocked and can go no further. DNA polymerase adds an A opposite the thymine present in the dimer. However, because of the distortion of the double helix in the region of the dimer, the DNA polymerase proofreads and "edits out" the newly-added A, which it mistakes for an incorrect base. This process is repeated over and over again, and DNA replication becomes stalled. Since replication is stopped until *something* is done, the SOS repair system allows the polymerase to "lower its standards," by relaxing the proofreading function of pol III. DNA polymerase is then able to continue replication *across* the dimer. This allows time for the real repair mechanisms to remove the dimers without the concomitant delay of DNA replication. (The synthesis of a number of the enzymes involved in these other repair mechanisms are simultaneously and coordinately derepressed when the SOS response is initiated!) The problem with this system is that it makes the polymerase "sloppy." Incorrect bases that are mistakenly incorporated into the newly-synthesized DNA molecule during the SOS response will remain uncorrected because proofreading is temporarily "turned off". For obvious reasons, this process is also called "error-prone replication."

Key Terms

Mutant

Mutation

Mutagen

Point mutation

Substitution

Insertion

Deletion

Temperature-sensitive mutation

Conditional mutation

Missense mutation

Nonsense mutation

Silent mutation

Leaky mutation

Transition mutation

Transversion mutation

Intercalation

Frameshift

Transposition

Polar mutation

Mismatch repair

Deamination

Depurination

Hot spot

Reversion

Revertant

Same-site reversion

Second-site reversion

Intragenic suppression

Intergenic suppression

Nonsense suppressor tRNA

Ames test

SOS repair

Excision repair

Recombinational repair

Postdimer initiation

Transdimer synthesis

Error-prone replication

Sister-strand exchange

AP endonuclease

Uracil-N-glycosylase

For More Information . . .

Where can I obtain additional explanations on these topics?

Mutation rates and examples of some base changes are briefly described on pp. 242-245 in Alberts *et. al.*, *Molecular Biology of the Cell*, 3rd edition.

The Molecular Biology of the Gene, 4th edition, by Watson *et. al.*, devotes much of Chapter 12 (pp. 339-345; 355-356) to the subject of mutagenesis.

Mismatch repair is described, with the aid of an elaborate diagram, on pp. 258-259 of *Molecular Biology of the Cell*, 3rd edition, Alberts *et. al.* SOS repair is briefly described on p. 249 of the same text.

DNA repair mechanisms are described in detail on pp. 350-354 of *Molecular Biology of the Gene*, 4th edition, Watson *et. al.* Page 350 contains an especially detailed illustration of excision repair and p. 351 illustrates a model that explains the details of mismatch repair.

The Wise Owl Says

What separates the molecular biologist from the nonscientist is the unrelenting search, by the molecular biologist, for cause and effect relationships. The main question that drives the molecular biologist is: "Why are things the way they are?"

Often, however, answering this question gets complicated. Let's take the example of the Rec A protein mentioned in Chapter 10 of *EMB.* Rec A is involved in the SOS repair system, but it has other functions as well. Some of those other activities are not directly involved in DNA repair. For example, when Rec A is activated by ultraviolet light, it in turn activates other proteins, including proteolytic enzymes, enzymes that cut proteins (a.k.a. "proteases"). These proteolytic enzymes in turn activate or inactivate other proteins.

In other words, Rec A triggers a cascade of changes. Despite the inherent complexity of this cascade, molecular biologists continue to work hard to sort it all out. In their relentless search for cause and effect relationships, it is sometimes discovered that cause becomes effect, and that effect becomes cause. Rec A is a case in point. UV *causes* Rec A to be activated. Rec A, an effect, then becomes a cause by activating proteolytic enzymes—an effect, that in turn become a cause by...

And so it goes!

Exercise

I. Understanding the effects of the chemical mutagens presented in Chapter 10 can be a difficult task. Approaching it in the following manner may be helpful. Here is a brief summary of the direct effect of each mutagen.

Nitrous Acid:	Changes C to U and A to H (H acts like G)
Hydroxylamine:	Changes C to hydroxylaminocytosine (which acts like T)
EMS:	Changes A*T to G*C and vice versa (transitions)
5-BU:	Changes A*T to G*C and vice versa (transitions)

Now, draw a cell with a short (3 bases) segment of double-stranded DNA. Pretend to expose the cell to a mutagen. Then redraw the cell and DNA molecule showing only the direct effect of the mutagen (i.e., a single-base change; not a single base-*pair* change). Now imagine that the DNA molecule replicates. Draw the two daughter cells and their DNA molecules. You should find that one of the two cells has a "normal" DNA molecule, while the other contains a mutation.

Compare your diagram with Figure 10-6 in *EMB*.

II. Let's review how hot spots increase the frequency of mutation. Please recall that the deamination of cytosine yields uracil. DNA-editing enzymes recognize any U, remove it, and replace it with C. Using the diagrams below, illustrate this process.

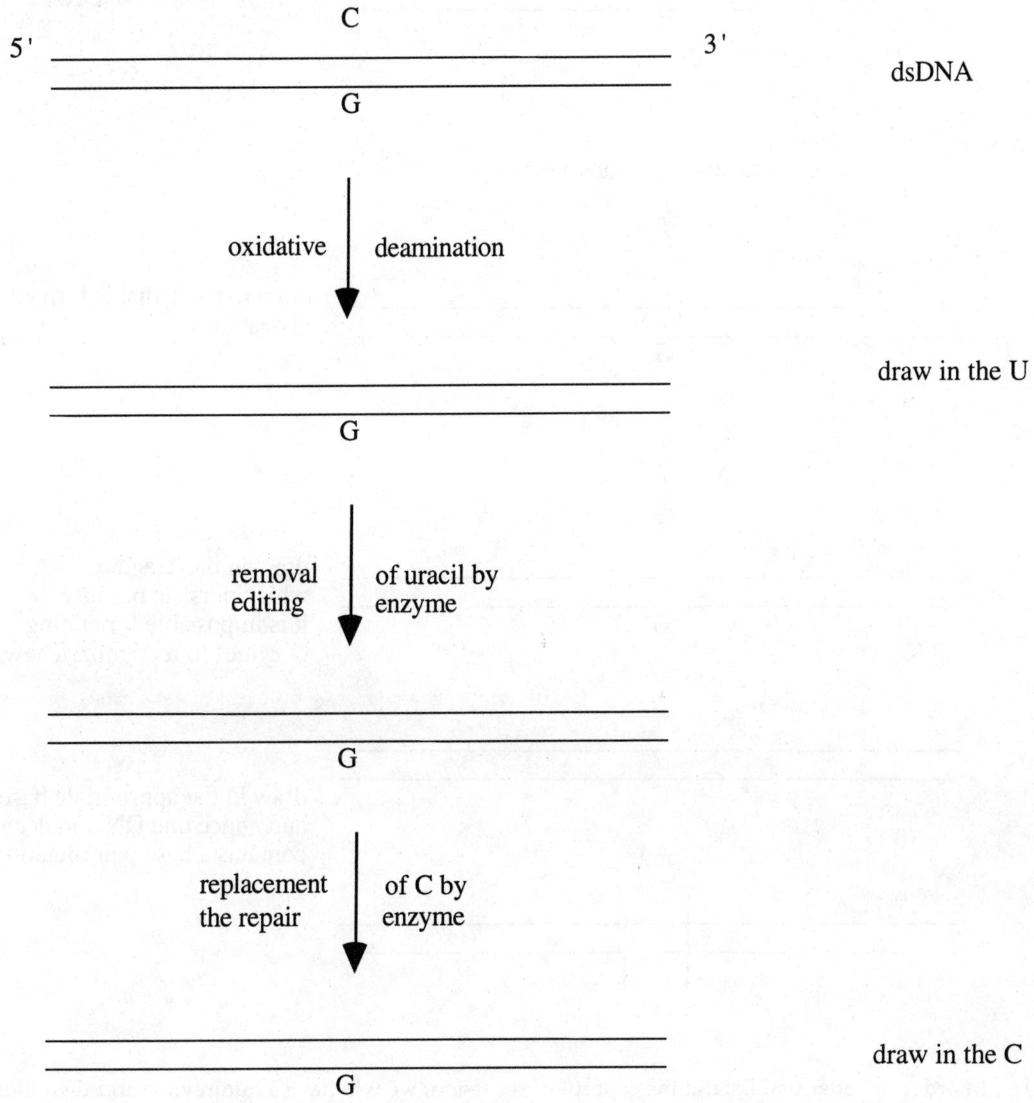

As you can see, the removal and replacement was straightforward because editing enzymes can easily recognize uracil in DNA; it does not belong there!

For methylated cytosine, often found in eukaryotic DNA, the process is more complex:

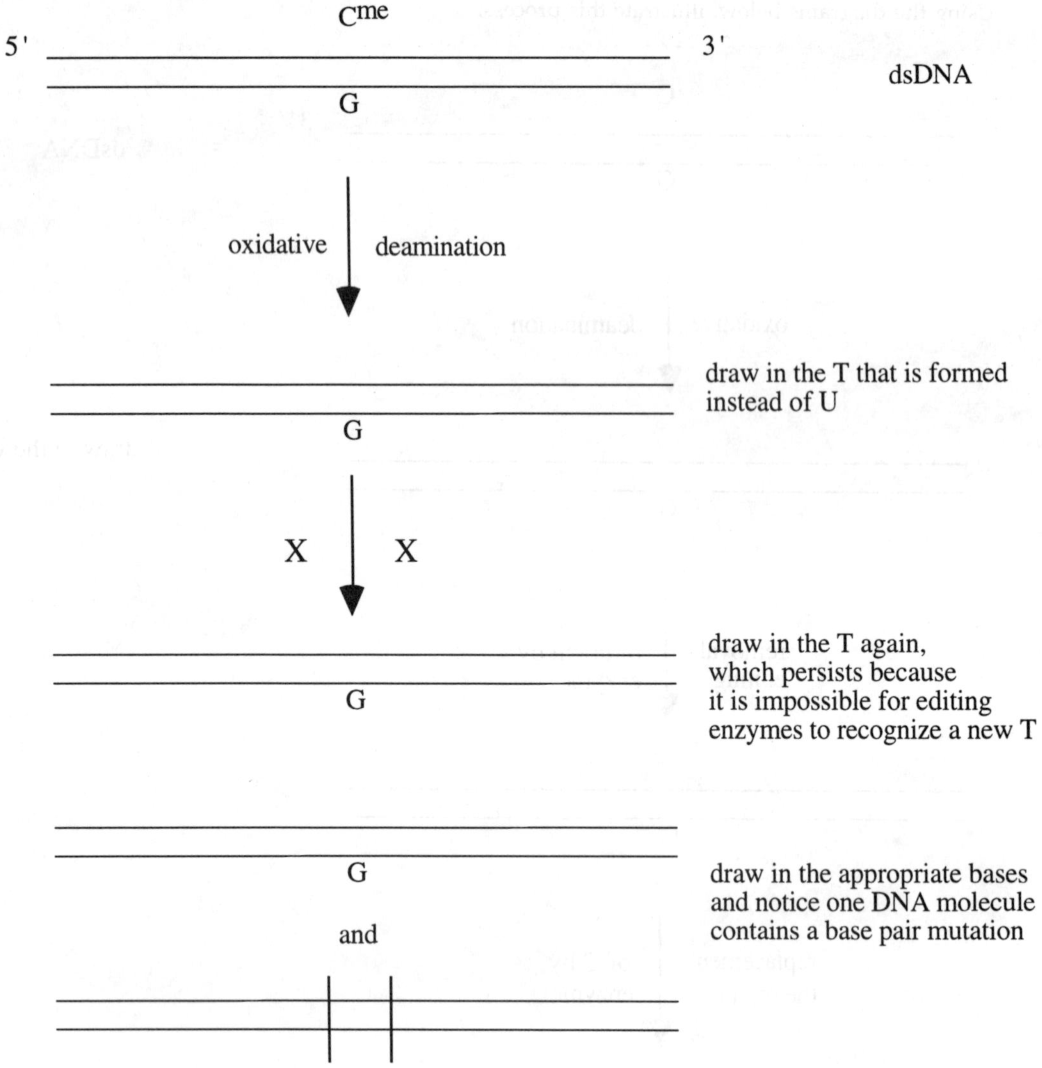

Cme
5'
3'
dsDNA
G
oxidative
deamination
draw in the T that is formed instead of U
G
X
X
draw in the T again, which persists because it is impossible for editing enzymes to recognize a new T
G
G
draw in the appropriate bases and notice one DNA molecule contains a base pair mutation
and

III. In order to better understand the concept of reversion, we will have a minireview and then illustrate the concept with a few diagrams that expand on some of the features shown in Figure 10-12 in *EMB*.

Reversions are sometimes called "back mutations." An exact reversal of the base change that comprised the original mutation is referred to as a *true* reversion. Naturally, this is the simplest type of reversion, or back mutation. Let's illustrate a true reversion at the level of protein folding.

Once completed, compare your drawing on the next page with Figure 10-12 in *EMB*. To which diagrams in that figure does this exercise pertain?

Original mRNA

(a) 

mRNA

translation and folding

mutation

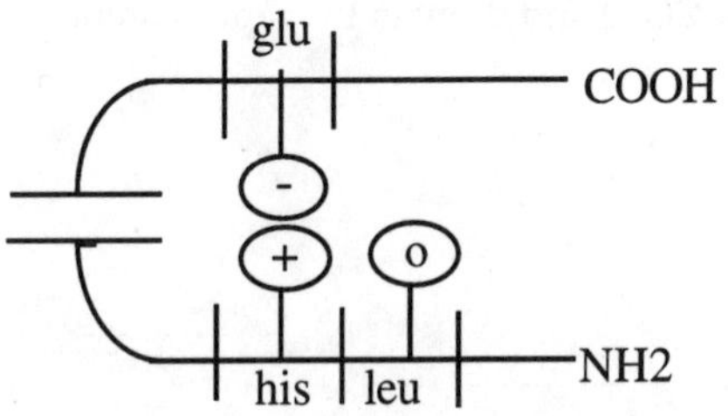

folded enzyme with full activity

Mutated mRNA

(b) 

mRNA

translation

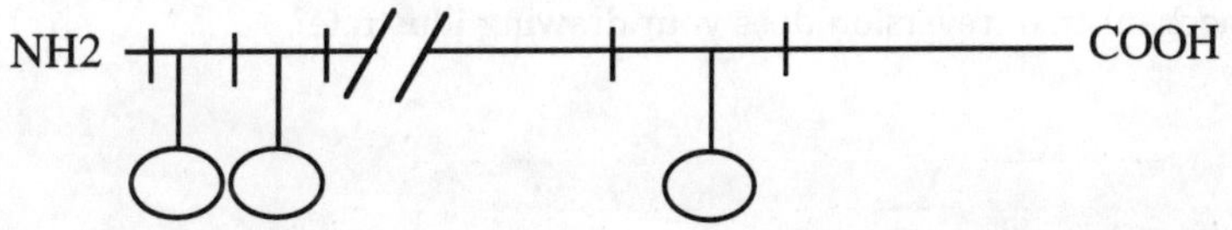

draw in the new amino acids using information provided in Table 9-1 in *EMB*

draw the folded protein, illustrating the effect of the amino acid substitution:

Reverted mRNA

(c) 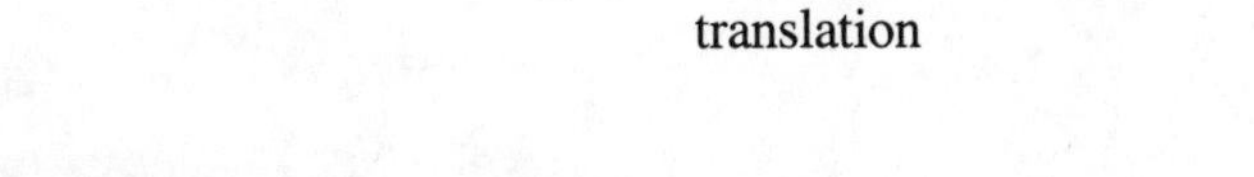

mRNA

translation

What shape will the protein display?

Occasionally, mutations occur elsewhere in the gene and correct the original mutation. These are called *second-site reversions* (or sometimes "intragenic suppression"). They correct, or compensate for, the original defect.

In order to better understand the concept of second-site reversion, return to the diagrams on the previous page.

Suppose a mutation occurred at the first codon and changed the codon from CUA to CGA. Look in Table 9-1 in *EMB* to find the amino acid substitution. On the original diagram, (b), illustrate the codon change (CUA → CGA) and draw in the replacement amino acid. Now, draw the new folded protein below:

How does this new shape compare with that of the folded protein in diagram (a)? Has this mutation reversed the effect of the original mutation? Compare your diagram with Figure 10-12 in *EMB*. Which mechanism of reversion does your drawing illustrate?

Reversions are sometimes called *back-mutations*. Can you explain why?

IV. UV reactivation workout!

Our goal is to understand how the data in Figure 10-18 in *EMB* generate the conclusion that a low dose of UV activates a "DNA repair" system.

First, let's understand that three separate experimental manipulations are involved. In the first manipulation, a sample of λ phage is irradiated at **several UV** doses. This is done by subdividing the phage sample into several small batches. Each batch is given a different UV dose. Some batches are given low doses, while other batches are given medium or large doses. The abscissa on the Figure 10-18 illustration refers to the UV doses given to the phage.

Second, two samples of host bacteria are prepared. One sample serves as a "control" and is not irradiated. In Figure 10-18 this sample is referred to as "unirradiated bacterial host." The other sample is "lightly irradiated."

Third, the samples of phage exposed to UV are plated on bacterial lawns. One set of plates contains a lawn of unirradiated host bacteria. Another set of plates contains a lawn of lightly irradiated host bacteria.

After the plaques form, they are counted. The percent of irradiated phage that survived on unirradiated host bacteria is then compared to the percent that survived on irradiated host bacteria: hence, the *two* survival curves in Figure 10-18. Clearly, regardless of the dose of UV given to the phage, the lightly irradiated host bacteria provide higher survival percentages.

Why? Presumably because a low dose of UV activates the repair system. The repair system helps repair the host DNA as well as the DNA of the infecting virus. Thus, the host's repair system enhances the survival of the irradiated virus. That is, it is the bacterial host's enzymes which repair the infecting viruses' UV damage.

V. Understanding mismatched bases.

To help understand how errors might occur during DNA replication, we will draw a mismatched base pair. Let's focus on the isomerization of cytosine because it is a relatively simple case.

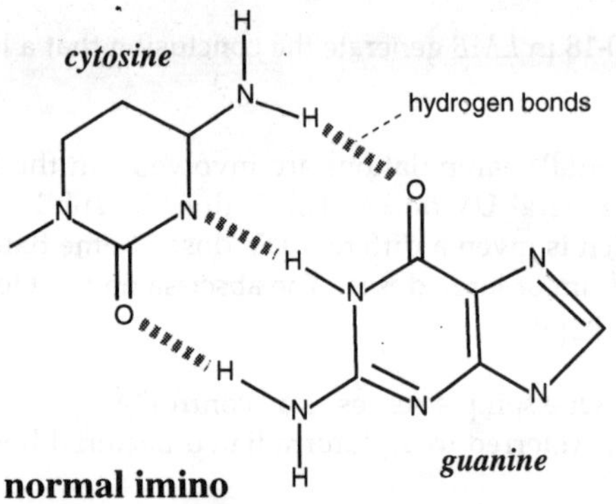

normal imino tautomeric form

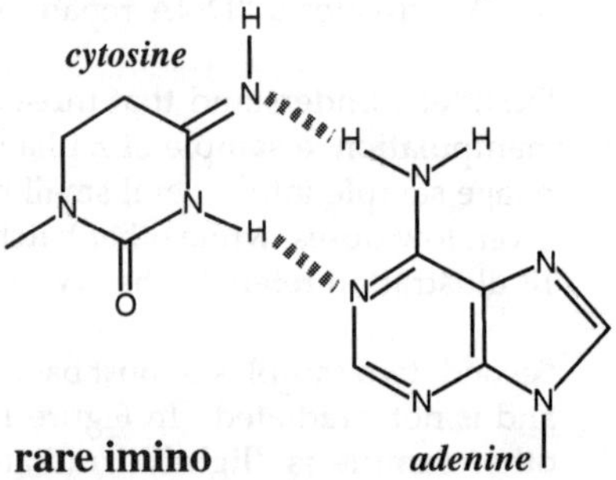

rare imino tautomeric form

In rare instances, C isomerizes, for a brief moment, just as DNA replication is occurring. Normally, of course, C pairs with guanine (G). The isomer above will, however, pair with adenine (A), rather than G. This results in a C-A mismatch in place of the normal C-G base pair.

For practice, draw in the structural formula of A and illustrate how it forms two hydrogen bonds with C.

Study Questions

1. Let's assume that you are studying a protein made up of 232 amino acids. Suppose you notice that the protein loses all function when a specific mutation occurs in amino acid 122, changing it from lysine to aspartic acid. However, function is restored when an additional mutation occurs, changing glutamic acid, normally at position 194, to glutamine. What can you say about amino acids 122 and 194 in the normal protein?

2. Which codons could become nonsense codons by changing only one base?

3. In a protein, what types of amino acid changes would be least likely to affect the function of that protein?

4. Explain how the following mutagens cause mutations: nitrous acid, 5-bromouracil, hydroxyl amine, and ethylmethane sulfonate. Are these transition or transversion mutations?

5. Suppose you have two strains of his^- bacteria. One strain contains a point mutation and the other contains a deletion. Devise a way to differentiate between the two strains.

6. Could an intergenic reversion be a same-site reversion?

7. Could 5-bromouracil cause a missense mutation by replacing methionine with valine? With threonine?

8. a. A point mutation is induced in a cell by exposure to hydroxylamine. Can the mutation be reverted by re-exposure to hydroxylamine?
   b. A point mutation is induced with nitrous acid, could it be reverted by reexposure?
   c. Can the mutation that was caused by hydroxylamine in part a. be reverted by exposure to nitrous acid?

9. What factors make the SOS repair system different from the other repair systems discussed in this chapter?

10. Describe how the mismatch repair mechanism can distinguish which strand of the DNA contains the incorrect nucleotide.

11. Why is recombinational bypass repair sometimes also called "sister-strand exchange"?

12. When is "sister-strand exchange" employed?

Answers to Chapter 10 Study Questions

1. Function is lost by changing the amino acid at position 122 from one with a positive charge (lysine) to one with a negative charge (aspartic acid). However, this is compensated for by changing the amino acid at position 194 from one with a negative charge (glutamic acid) to one with a positive charge (glutamine). Function of this enzyme must, therefore, depend in part on a conformation induced by ionic bonding between positions 122 and 194.

2. UUA, UUG, UCA, UCG, UAU, UAC, UGU, UGC, UGG, CAA, CAG, CGA, AAA, AAG, AGA, GAA, GAG, and GGA.

3. If the mutation does not cause a change in charge, sign, polarity, or other chemical properties, then there is a good chance that function will not be affected. Also, if the amino acid change occurs in an area of the protein that is not important to its function, an effect may not appear.

4. Nitrous acid causes deamination of adenine to hypoxanthine (which pairs with cytosine), cytosine to uracil (which pairs with adenine), and guanine to xanthine (which pairs with cytosine). 5-bromouracil acts as an analog of thymine and is incorporated across from an adenine. It can also tautomerize; when it does it can pair with guanine. Hydroxylamine acts specifically on cytosine, causing it to base-pair with adenine. EMS is an alkylating agent that modifies guanine, causing it to pair with normal thymine and modifies thymine, causing it to pair with normal guanine. These mutations caused by nitrous acid, hydroxylamine, and EMS are transition mutations. 5BU mutations are always transitions. Nitrous acid and EMS, however, can sometimes cause transversions.

5. Expose some colonies from each strain to mutagens that induce point mutations. Revertants should result in the strain that contains the point mutation but not in the strain with the deletion.

6. No, this is a matter of definition, or vocabulary, but it is an important point nonetheless. In an intergenic reversion, a reverting mutation occurs in a different gene than the one that contains the original mutation.

7. Yes, by replacing the thymine in the template strand of the DNA molecule. Again, yes, this time by replacing the thymine in the nontemplate strand of the DNA molecule.

8. a. No, hydroxylamine reacts only with cytosine, converting it to something that pairs only with A, thereby inducing a CG -> TA transition. Since it reacts only with C, it can not cause the reverse.
   b. Yes, nitrous acid causes both AT -> GC and GC -> AT transitions.
   c. Yes, nitrous acid does cause an AT -> GC transition.

9. First, the SOS repair system is activated only when needed, contrary to most other repair systems. Second, it does not actually repair damaged DNA. The DNA it synthesizes needs to be further repaired by other "DNA repair systems."

10. The mismatch repair system can recognize the degree to which each DNA strand is methylated. The parental strand is usually highly methylated, whereas the newly synthesized daughter strand is not. If the nucleotide is not methylated, then it most likely belongs to the daughter strand, and is thereby recognized as the incorrect nucleotide.

11. Because in the recombinational bypass mechanism a single-stranded segment is excised from a "good" strand of DNA and "recombined" into the gap where a defective region has been excised.

12. Sister-strand exchange is employed to patch the gap created when a region containing a thymine dimer is bypassed during replication.

Concept Map

A concept diagram will help us understand the diverse sections of this chapter. Let's prepare one!

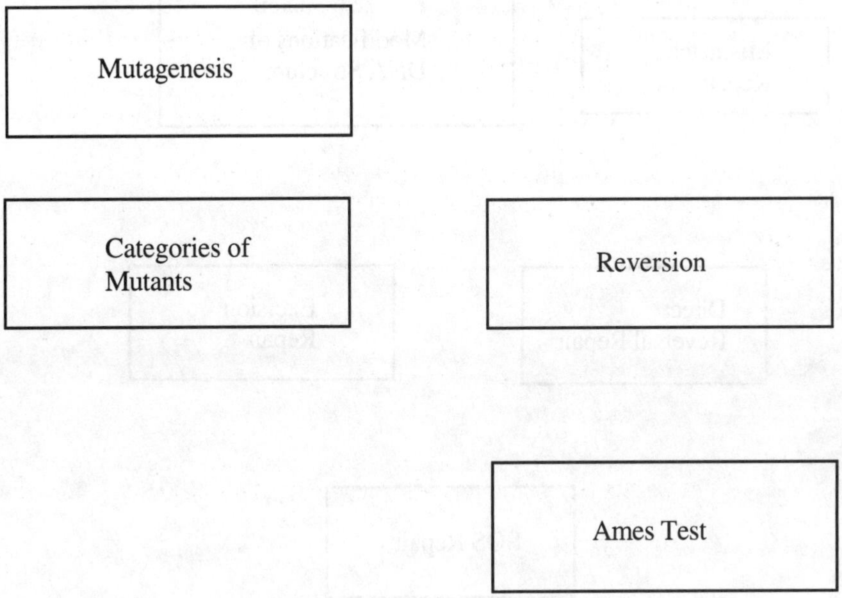

To insure that your concept map is complete, please include the following terms. Also, scan the "Key Terms" section and include as many of those terms as you feel necessary.

- spontaneous mutant
- leaky mutant
- hot spots
- phenotype
- genotype
- induced mutant
- transposable elements

Concept Map

DNA repair mechanisms might be better understood if we relate them to each other with a concept map:

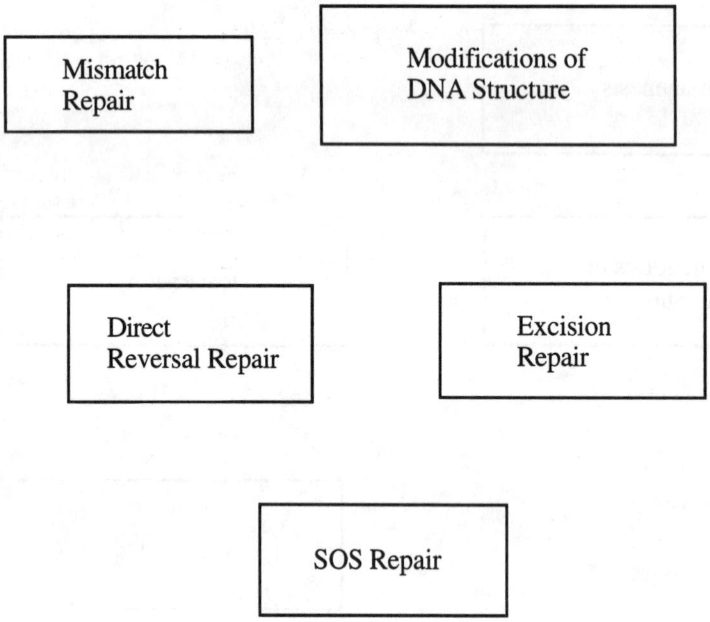

Let's be sure we include the following terms in our diagram:

- deamination
- photolyase
- postreplicational repair
- single-strand breaks
- depurination
- recombinational repair
- error-prone replication
- cross-linking

Science and Society Issues

Cancer is generally viewed by molecular biologists as representing a mutation in a cell or tissue type that eventually leads to uncontrolled growth, or proliferation of that cell or tissue. Most carcinogens have been demonstrated to have potent mutagenic activity. In several instances, substances that initially lack mutagenic activity are converted by the living cell's metabolic enzymes into potent carcinogens.

Many scientists have expressed deep concern over the proliferation of novel chemical compounds, many of which are potential carcinogens and some of which are being released either into the environment as industrial waste, or in food as additives for enhancing flavor, extending shelf life, or improving aesthetic appeal.

Since approximately 20% of the population of highly developed countries already dies from cancer, increasing concern about environmental carcinogens is justified. Most cancers do not arise as the result of a single mutation. Instead, several mutations must exert their effects. These effects may culminate in the loss of normal growth content by a cell or tissue.

How can we convert the information about mutagens and carcinogens, developed in Chapter 10 of *EMB*, into social policy? Is it a scientific issue or a social issue? What role should scientists play in establishing public policy?

Practical Applications

The study of mutations and mutagenic agents has provided an enormous amount of insight into virtually all aspects of molecular biology. As mentioned in *EMB*, the analysis of mutant phenotypes provides invaluable clues as to the function of various molecular mechanisms in the living cell.

In addition, practical applications have emerged from the study of mutagenesis. Genetic engineering depends heavily on mutagenesis to attain the production of new and different types of enzymes. The introduction of novel amino acids into the primary sequence of a protein requires the alteration of the nucleotide sequence of the gene. Originally, mutagenic agents were directly applied to living cells to introduce mutations. Now, using the principles outlined in Chapter 10 of *EMB*, changes in nucleotide sequence are not usually made by mutating whole organisms. Rather, oligonucleotide synthesis procedures are employed to generate altered gene sequences. By the use of recombinant DNA techniques that you will be introduced to in Chapter 15, these altered genes are inserted into the appropriate gene in the organism and are eventually expressed. That is, proteins can be engineered by making specific modifications to relevant genes.

Another practical application of mutagenesis is the Ames test, described in Chapter 10 of *EMB*. A special strain of the bacterium *Salmonella* is employed as a test organism in this test. This strain is unable to synthesize the amino acid histidine. It has also been genetically altered so that its DNA repair system is defective. It is, therefore, especially sensitive, since it cannot easily repair new mutations.

Potential carcinogens are used to generate reversions of the His^- phenotype back to wild type (capable of synthesizing histidine). Since this strain can't easily repair mutations, the effectiveness of the carcinogen can be accurately monitored. The relative danger, in terms of cancer-causing potential, of various chemicals can thereby be easily and inexpensively measured.

Molecular biologists have discovered that most cancers form as a result of an alteration in the DNA sequence of a cell. Most carcinogens are, therefore, mutagens. Hence, the use of mutation (i.e., reversion) analyses as the basis for predicting the cancer-causing potential of a chemical substance.

11 Regulation of Gene Activity in Prokaryotes

Here's Help

What does regulation mean, anyway?

In general, regulation means "the control of the amount of a gene product." Most of the regulatory (i.e., controlling) mechanisms described in Chapter 11 of *EMB* act during transcription by increasing or decreasing the amount of mRNA synthesized from a specific gene. Several examples of controls that act at the level of the translation (protein product) are also described toward the end of Chapter 11.

Why is regulation important to cells?

During the life cycle of a cell the needed amount of each type of the few thousand proteins a cell contains varies. While growing rapidly, it needs enzymes involved in DNA replication. When confronted with novel environmental circumstances, it needs appropriate metabolic enzymes. In order to meet the changing needs of the cell, molecular mechanisms have evolved to increase or decrease the amount of individual gene products. The phenomenon of altering the quantity of specific gene products is referred to as "gene regulation, " or "gene control."

What is the difference between positive and negative control of transcription?

Both positive and negative control involve at least one type of regulatory element. In positive control, the regulatory element activates transcription, which will not occur unless that element *is present*. In negative control, the regulatory element, when present and active, prevents transcription of the operon. In this latter case, transcription can only occur if that element *is removed* from the operon.

How about an overview of the data in Table 11-1?

The genetic combinations illustrated in that table are of two general types: genotypes #1, 2, and 5 represent bacterial strains alone; all the other genotypes represent bacterial strains into which extra genes, carried by a plasmid, have been introduced. Thus, these "other" strains are actually "partial diploids".

These genetic combinations are very useful for deducing general features of gene regulation. For example, let's see what types of information can be deduced by looking at partial diploids #3 and #4.

Strain #3 genotype: **F'** $I^-\ O^+\ lacZ^+/I^+\ O^+\ lacZ^{+ (or -)}$

Strain #4 genotype: **F'** $I^+\ O^+\ lacZ^+/I^-\ O^+\ lacZ^+$

The major difference between the two strains is the location of functional *lac* repressor gene ($lac\ I^+$). In #3, it is situated on the bacterial chromosome, in proximity to the *lacZ* gene; in #4, the functional copy of the repressor gene is located on the F ' plasmid. Nevertheless, neither strain exhibits *lacZ* gene expression in the absence of the inducer IPTG, while both *do* produce the *lacZ* gene product in the *presence* of the inducer. This tells us that the *lacI* gene does not need to be situated next to the genes it controls, and that its gene product is likely to be a soluble, diffusible protein.

Let's look at another example involving strains #6 and #7.

Strain #6 genotype: **F'** $I^+\ O^c\ lacZ^+/I^+ O^+\ lacZ^-$

Strain #7 genotype: **F'** $I^+\ O^c\ lacZ^-\ I^+ O^+\ lacZ^+$

In strain #6, the mutant operator O^c is located next to the functional $lacZ^+$ gene on the F ' plasmid, while the functional operator O^+ is found next to the mutant $lacZ^-$ gene on the bacterial chromosome. Expression of the *lacZ* gene is constitutive in this instance. In strain #7, the mutant operator O^c is located next to the non-functional *lacZ*$^{-}$ gene on the F ' plasmid, while the bacterial chromosome contains the functional operator situated adjacent to the functional *lacZ*$^{+}$ gene. In this instance, expression of the *lacZ* is inducible. Looking at both of these sets of data together, it appears that the *lacO* gene (operator) must be located adjacent the genes that it controls, unlike the *lacI* gene which can act at a distance.

Using this genetic approach, it was possible to develop a conceptual framework for searches for the actual genes and molecules which comprise the regulatory system for lactose metabolism.

Here's a brief review of the *lac* operon.

The *lac* operon contains the cistrons that encode the necessary enzymes for the metabolism (breakdown) of the sugar lactose. This operon is regulated such that the enzymes will not be produced unless two requirements are met: The cell must be growing in the *presence* of lactose **and** the *absence* of glucose. There is a good reason for this: The cell prefers glucose as a carbon source. When glucose is present, the cell has no need to use lactose, so the *lac* operon should be "off." However, when no glucose is available but lactose is present, the cell can either use lactose, or starve (unless, of course, other sources of carbon were available). In this situation, the lactose operon should be "on." How does the cell know how much glucose is available? A molecule called cyclic AMP (cAMP) (see Figure 11-7 of *EMB*) is present in large amounts whenever glucose is not. One of several requirements necessary for transcription of the *lac* operon is for a protein (CRP) complexed with cAMP to bind to the operon. Transcription *cannot* occur unless this CRP-cAMP complex is bound to the operon (binding occurs at the activation site within the promoter region). When glucose is present cAMP is not. Therefore, there will be no CRP-cAMP complex to bind to the operon and allow transcription when glucose is present. Sound simple? Unfortunately, we have to keep in mind the effects of more than just glucose. Remember, the cell only wants to transcribe the *lac* operon when needed: that is, when there is no glucose present and lactose is available.

Located between the promoter (where RNA polymerase binds to initiate transcription) and the first *lac* gene, is a region called the operator. Please see Figure 11-6 of *EMB* for an illustration of the structure of the *lac* operon. A gene called *lacI* makes a protein molecule that is present in the cytoplasm and can bind to this operator region. Upon binding, this protein effectively acts as a "roadblock" for RNA polymerase and no transcription can occur. This is where lactose comes into play. An isomer of lactose, called allolactose, binds to these repressor proteins and prevents them from binding to the operator when lactose is present (see Figure 11-6(b)). Therefore, in the presence of lactose, no repressor will be bound to the operator and RNA polymerase will not be blocked. Provided that glucose is not present, so that the cAMP-CRP complex is bound to the operon, transcription can occur.

Now, let's examine how the *lac* operon functions under different growth conditions. If a cell is growing only in glucose, it has no reason to express the genes in the *lac* operon. Very small amounts of cAMP will discourage cAMP and CRP from complexing and binding to the operon to stimulate transcription. Also, since lactose is not present, the repressor will be free to bind to the operator and block transcription. The opposite situation exists when a cell is growing only in lactose. The lactose keeps the repressor from binding to the operator and interfering with transcription, while the high level of cAMP (due to the low level of glucose) means that there will be cAMP-CRP complexes bound to the operon, driving transcription.

Suppose a cell is growing in glucose and lactose. Since lactose is present, no repressor should be bound to the operator and transcription should occur—right? Wrong! It's true that no repressor will block transcription, but that isn't the only factor to be considered. Since glucose is present, cAMP is not. Therefore, no cAMP-CRP complex will be bound to the operon, and without this, little transcription will occur. Just because there is no repressor doesn't mean that transcription can occur.

Let's now imagine a situation in which a cell is growing in medium that contains neither glucose nor lactose. The concentration of cAMP will be high, so there will be plenty of cAMP-CRP to bind to the operon, conducive to transcription. However, there will also be plenty of repressor available to bind to the operon; this will prevent transcription.

What happens when the various regions of the *lac* operon are mutated?

Let's refresh our memory about the various regions of the operon. The regions associated with the operon that are discussed in Chapter 11 of *EMB* are the gene that encodes the repressor protein, the promoter, the operator, and the genes for ß-galactosidase and lactose permease.

If the gene that encodes lactose permease were mutated, the cell would not be able to transport lactose into the cell, so the cell will be Lac⁻, even though lactose could be metabolized if artificially introduced into the cell. Note that the cell could be "fixed" by bringing in a functional copy of this gene on a plasmid, as there would then be permease present.

The same is true for the galactosidase gene. If it were mutated such that no functional galactosidase could be made, the cell would also be Lac⁻. This cell would be able to transport lactose in, because the gene for permease would still be functional, but that lactose would not be metabolized. If a normal copy of the galactosidase gene were brought into the cell on a plasmid, then functional galactosidase would be present and the cell would be able to metabolize lactose.

What do you suppose would happen if the promoter were mutated? Well, the promoter is the site to which RNA polymerase binds to initiate transcription. If it were so extensively mutated that the enzyme couldn't recognize it, then there would be no mRNA synthesis and, therefore, neither permease nor galactosidase would be produced. Again, this cell would be Lac⁻. We recognized in the last two examples that bringing in a functional copy of the mutated region of the operon on a plasmid would revert the cell to Lac⁺. Would the same "trick" work in this situation? If a functional *lac* promoter on a plasmid were brought in, it would bind RNA polymerase. Nevertheless, this wouldn't benefit the cell, since the polymerase wouldn't be anywhere near the operon, but would be on the plasmid. Hence, reversion by this method would not be possible.

Another important region is the operator region. The repressor protein binds here and acts as a roadblock to RNA polymerase. If the operator were deleted, the active repressor would have nowhere to bind, so transcription would never be repressed, not even when lactose is absent. Again, bringing in a functional operator on a plasmid would do no good, as the repressors would bind to the operator on the plasmid, which would not block the way of the polymerase.

What would happen if no repressor proteins were made? Well, this would have the same effect as an operon without an operator; repression would not be possible. In this case, however, the addition of a functional copy of the gene on a plasmid would correct the cell. The repressor would then be made and could bind to the operator as usual, in the absence of lactose.

What is DNA looping?

The image of DNA developed in Chapter 3 of *EMB* (e.g., Figure 3-1) suggests that DNA is a relatively linear (straight line) polymer of nucleotides. In Chapter 7, however, DNA was shown to form a loop at the replication fork (Figure 7-22). The study of gene regulation has revealed additional instances of "looping," or folding, of double stranded DNA. Figure 11-12 of *EMB* illustrates this looping.

A double-stranded DNA molecule can bend into a loop over approximately 6 turns of the helix. That is, approximately 60 bp of dsDNA is a sufficient length to permit a loop to form. Within a loop, a DNA binding protein can simultaneously bind to both arms of the loop. By keeping the dsDNA in a looped configuration, the DNA binding protein (e.g., a specific gene repressor protein) can block transcription of an adjacent gene.

How does the *trp* operon described in Figures 11-13 and 11-14 of *EMB* work?

As you already know, the *trp* operon contains the cistrons that code for the enzymes that synthesize the amino acid tryptophan. (Please note: the *trp* operon does not itself synthesize the amino acid tryptophan. Rather, this operon codes for the various enzymes which themselves comprise the metabolic pathway that is responsible for tryptophan synthesis.) As with many operons, when its services are not needed it benefits the cell to be turned off. In other words, when tryptophan is present in large amounts, the cell will use what it already has instead of making more (i.e., the operon will be turned off). When the level of tryptophan is extremely low, however, the cell will need to make its own. Thus, the genes that code for the necessary enzymes will be expressed. To handle the "all-or-none" situations (you must bear in mind these situations probably do not really exist in nature, but thinking about them in this way makes things simpler), there is a repressor molecule that prevents transcription when bound to the operon. This repressor molecule binds to the operator only when the level of tryptophan is high. In order for the repressor to be able to bind the operon, it first must be complexed with a molecule of tryptophan. Therefore, in low levels of tryptophan there will be little active repressor, and transcription will occur. In high levels there will be a lot of active repressor (because a lot of tryptophan is present), so there will be little transcription. Neat, eh!

In addition to the repressor that prevents transcription at high levels of tryptophan, a second method of regulation for expression of the *trp* operon genes exists. This mechanism serves to "fine tune" *trp* expression. When a transcript is synthesized, it contains a region at the 5' end (the end where the ribosome starts translating) called the "attenuator." If the level of tryptophan is low enough that transcription occurred (because not much active repressor was formed), but there is still too much tryptophan in the cell, then transcription of the operon will be prematurely terminated. In other words, the RNA polymerase will be stopped at this attenuator sequence before it transcribes any of the genes coding for the tryptophan synthesis enzymes. This attenuator sequence consists of four regions, called regions 1, 2, 3, and 4. They are illustrated in Figure 11-14 in *EMB*. (Region 1 is at the 5' end, followed by 2, 3, and 4, respectively). These regions have several important properties: Region 1 has two tryptophan codons on its 5' side, while regions 3 and 4 are able to base-pair with each other. When regions 3 and 4 base-pair, they form a stem-loop structure that acts as a Rho-independent terminator of transcription (i.e., a stem-loop followed by several uracils). This structure is diagrammed on the right-hand end of Figure 11-14(c). If this structure is allowed to form, then RNA polymerase is stopped before it transcribes any of the structural genes. Region 2 can also form a stem-loop structure by pairing with region 3. The 2-3 stem-loop is, however, *not* a terminator, because it is not followed by a series of uracils. You can see this in Figure 11-14(d).

How do all these properties interact to regulate translation of the transcript? Well, imagine the ribosome going down the mRNA molecule. As it comes upon each codon, the appropriate tRNA adds the appropriate amino acid. A problem may arise when it gets to the two tryptophan codons at the beginning of region 1. If there is not much tryptophan, then there will not be much charged tryptophan tRNA. Since there is not much tryptophan tRNA, the ribosome will be stalled "on top" of region 1 while it waits for tryptophan tRNA. As long as it is stalled on region 1, nothing prevents regions 2 and 3 from forming a stem-loop. The key idea here is that if the 2-3 stem-loop forms, then the 3-4 stem-loop obviously *cannot* form. Since this 3-4 stem-loop does not form, there is nothing to prevent the polymerase from transcribing the downstream genes. This is good, because if there wasn't even enough tryptophan in the cell to charge a tryptophan tRNA, then the cell certainly needs to make some.

Now, suppose the ribosome had begun translating the transcript, but the cell did not really need to make tryptophan. How will transcription be stopped before energy is wasted making unnecessary mRNA and enzymes? Well, since there is sufficient tryptophan, the ribosome will not stall on top of region 1. Since the ribosome will not be stalled, it will move "on top" of region 2, which will obviously prevent the formation of the 2-3 stem-loop. Region 3 will then pair with 4, since it can't pair with 2, forming the 3-4 stem-loop. This structure is a Rho-independent terminator and will force the polymerase to stop transcription before reaching any of the structural genes. Remember, this only occurs when synthesis of tryptophan is not needed.

A word of caution!

Please be careful when comparing the regulation of the *trp* and *lac* operons. Remember, each set of enzymes works in an opposite pathway. The enzymes of the *lac* operon are involved in a degradative pathway; they break the substrate down when that substrate is present in the cell. Conversely, the enzymes of the *trp* operon are involved in a biosynthetic pathway; they make a product when it is *not* present in the cell. This is probably why lactose *inactivates* the repressor of the *lac* operon, but tryptophan *activates* the repressor of the *trp* operon.

In a nutshell, what does feedback inhibition (Figures 11-17 and 11-19) accomplish for the cell?

Feedback inhibition provides a "rapid response" mechanism whereby the cell can quickly adjust or "fine-tune" the flow of metabolites through a pathway. Suppose a cell encounters a sudden increase in the amount of a metabolite it was in the process of synthesizing itself. Almost instantly, it can slow down its own production of that metabolite by shutting down the activity of one of the key enzymes in its synthesis pathway. The cell thereby conserves the energy it would otherwise be expending for the synthesis of that metabolite.

Key Terms

Coordinate regulation

Negative regulation

Positive regulation

Repressor

Corepressor

Derepressor

Aporepressor

Inducer

Activator

Feedback inhibition

Lactose permease

lacZ

lacY

IPTG

Constitutive mutation

Operator

Lac operon

CRP

Attenuation

Allosteric proteins

Isoenzymes

For More Information . . .

Further explanations on these topics are indeed available:

The *lac* operon system is briefly described on pp. 419-420 of Alberts *et. al.*, *Molecular Biology of the Cell*, 3rd edition. DNA binding proteins are extensively described on pp. 404-421 of the same text.

The Molecular Biology of the Gene, by Watson *et. al.*, 4th edition, contains a whole chapter on the subject of gene function regulation in bacterial cells. See Chapter 16, pp. 463-502.

The Wise Owl Says

How much detail you see depends upon the power of your binoculars.

As mentioned in the beginning of Chapter 11 of *EMB* , bacteria and bacteriophages form the basis of all important present-day concepts of regulation, since theirs are the best understood regulatory mechanisms. These simple systems are well understood because powerful methods, such as fine structure genetic mapping, recombinant DNA technology, biochemical fractionation procedures, and other tools, can be used in remarkably sophisticated ways to understand the details of many molecular mechanisms.

The initial observations usually yield only the outlines, or general features, of a molecular process. As more powerful tools are employed, the details come into finer focus. The details of many prokaryotic molecular processes are well understood. This is, however, not the case for eukaryotic organisms. The amount of detailed knowledge associated with most eukaryotic molecular mechanisms is minimal by comparison with prokaryotic systems. This will become apparent when you read Chapter 13 of *EMB*.

Why is there such a discrepancy between our knowledge of prokaryotic and eukaryotic molecular systems? First, high powered genetic tools are not currently available for use in most eukaryotic organisms. Second, the "efficiency" argument, mentioned in Chapter 1 of *EMB* , does not pertain to eukaryotic organisms. Natural selection has not constrained the evolution of regulatory mechanisms in eukaryotes nearly as much as in prokaryotes. Hence, the regulatory mechanisms of eukaryotes are inherently more complex than those of prokaryotes.

As higher power tools, such as recombinant DNA technology and complete nucleotide sequences of genomes, become available for use in the study of eukaryotic mechanisms, they should enable a level of detail similar to that achieved in prokaryotes, and astonishing amounts of complexity will undoubtedly be observed.

Exercise

The following two exercises will help in learning how regulatory mechanisms in prokaryotic organisms work. Enjoy!

I. Let's work up a diagram that illustrates "DNA looping." Begin with the double-stranded DNA molecule shown below.

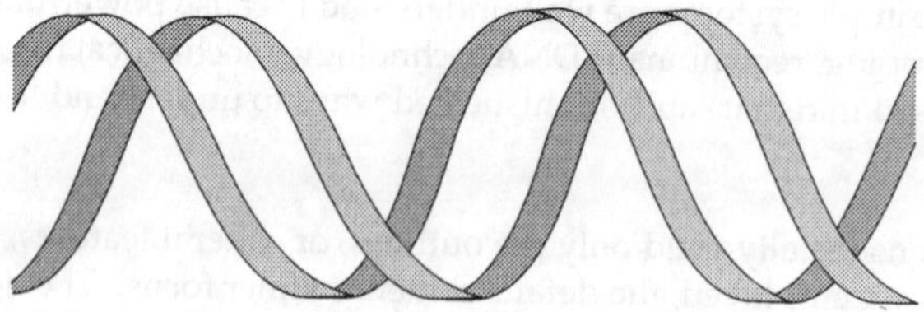

Our goal is to reconfigure it into a U-shaped loop, of the type shown in Figure 11-12b of *EMB*. Below, draw a loop that includes approximately 6 helical turns in the bend.

Now, add to your illustration how a dimer of a DNA binding protein might cross-link the 2 arms of the loop, thus stabilizing the loop:

II. Diagrams of hypothetical regulatory circuits can be very useful starting points in designing experiments. We will increase our understanding of both positive and negative control processes by preparing a circuit diagram. Our diagram will be versatile, as it will allow for both induction and repression modes of regulation.

In all cases, two separate genes (segments of the chromosome) are involved. The shorter segment on the left side of each box codes for either the inducer protein or the repressor protein, depending on the type of control designated for that box (positive or negative control, respectively).

Complete the circuit diagram on the following page and then compare it with the answer at the end of this chapter.

Positive Control Mechanisms	Negative Control Mechanisms
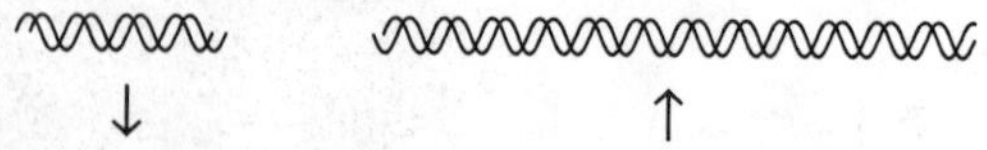 → 1. Draw in the transcription and translation of an inducer protein, from the left-hand arrow. 2. Illustrate how an activator molecule might activate the inducer protein (around the horizontal arrow). 3. Draw in the movement of activated inducer to the target sequence on the right-hand gene. INDUCTION MODE	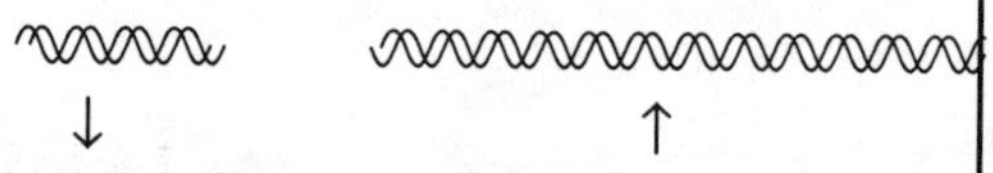 → 1. Draw in transcription-translation of a repressor protein, from the left-hand arrow. 2. Illustrate its inactivation by blocking protein (around horizontal arrow). 3. Diagram the consequences to gene expression, around the target gene. INDUCTION MODE
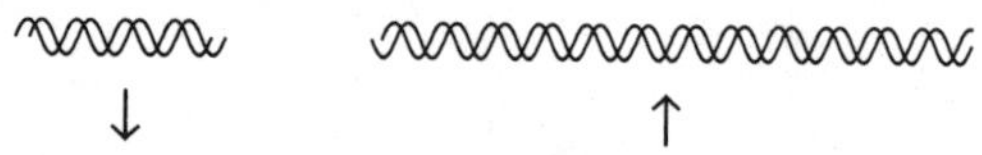 → 1. Draw in transcription-translation of an inducer protein, from the left-hand arrow. 2. Illustrate how a blocking molecule might inactivate the inducer protein (around the horizontal arrow). 3. Draw in the consequences to gene expression of inactivated inducer protein. REPRESSION MODE	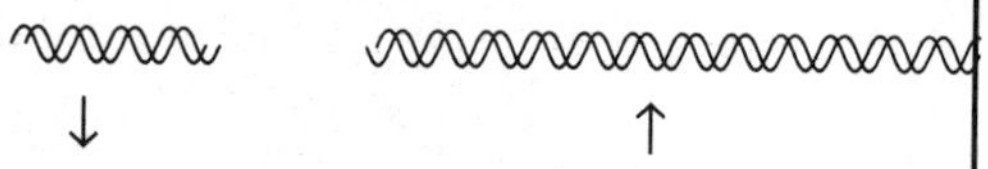 → 1. Draw in transcription/translation of a repressor protein, from the left-hand arrow. 2. Illustrate how an activator molecule might activate the repressor protein (around the horizontal arrow). 3. Draw in the consequences to gene expression of activated repressor protein. REPRESSION MODE

Tough Nuts

As the fine details of gene regulation in prokaryotic organisms are analyzed, several difficult-to-research issues have emerged. Here are two:

I. DNA looping provides a potentially complex issue for the molecular biologist who studies the regulation of gene expression. The interaction between proteins and DNA appears to comprise a key step in controlling gene expression in prokaryotic organisms (e.g., Figure 11-9 of *EMB*). In some instances, double-stranded DNA bends to form a loop. The DNA binding proteins help to stabilize the loop and thereby block transcription of surrounding genes.

Several questions arise when attempting to develop a sophisticated understanding of the looping phenomenon:

1. Do certain nucleotide sequences, such as A-T rich regions, facilitate bending of a DNA duplex?

2. How much energy is required to bend dsDNA, and where does that energy come from?

3. In the cytoplasm of a living cell (versus the molecular biologist's favorite venue for studying mechanisms—the test tube), DNA is probably in a dynamic state and is usually supercoiled. That is, various proteins, RNAs, and small molecules (e.g., physiologically important salt ions) are undoubtedly continuously interacting with the DNA. How then can we develop an *in vitro* (i.e., test tube) reconstruction of such a complex phenomenon, so that DNA can be studied in a relatively natural condition?

One approach has been to study examples of naturally curved or bent DNA fragments. Certain regions of bacteriophage l, for example, exhibit a natural curve and have been studied. One can ask, however, whether these ultra-simple systems provide suitable models for genes in their natural (living cell) environment.

II. How many control points or regulation steps exist for a specific gene? This sounds like a simple question; however, inventorying the control points associated with the regulation of a single gene appears, at times, to be an endless process. When one considers that controls can exist at three different levels: transcription, translation, and protein processing, this question takes on a more complex character; it becomes a tough nut.

The use of mutant genes (see, for example, Table 11-1 of *EMB*) has provided the main insight into how molecular mechanisms work; however, it has an inherent limitation. Let's first review the main strength of mutant gene research. Mutants that delete a function, or inactivate a process, reveal important information. For example, the normal form (allele) of a certain gene is found to be indispensable for a specific function or process, since the mutant version of the gene erases that function or process. Thereby, mutants provide an estimate of the minimal number of genes that are involved in a process.

There is a catch, however, which returns us to the "inherent limitation" alluded to above. Many mutations go undetected and are not recognized as mutant phenotypes by the molecular biologist. Why? Because although the function the genes specify may be involved in an expression event, they are not *indispensable*. The key word here is "indispensable." Only indispensable gene functions are detected by mutant analyses, since only they yield recognizable phenotypes.

Here we introduce the concept "gene redundancy." There are numerous instances wherein the loss of one gene product is compensated for by the presence of a substitute, or "pinch-hitter" gene product. Hence, mutant gene research compels scientists to study only indispensable features of a process or function.

At present, there is no way to predict which components of a process are protected by the existence of substitute mechanisms. Recall what the Wise Owl said in Chapter 6 about the predictive value of the science of molecular biology.

Certainly, the study of mutant genes has provided astonishing insights into regulatory mechanisms. The elucidation of the *lac* operon represents the first, and most dramatic, example of the power and achievement of a mutant gene approach.

Please keep in mind, however, that the mutant gene picture might be lopsided! The true picture of regulatory processes will probably involve redundancy. But how can we establish whether or not redundant mechanisms are involved in a specific process? That is:

What approach should be taken to crack these tough nuts?

Study Questions

1. In what two regions of the *lac* operon could a mutation result in the production of galactosidase and permease whether or not lactose is present?

2. How many different proteins are bound to the lactose operon in a cell growing in medium that contains neither lactose nor glucose? How many proteins are bound to the operon if only lactose is present? If only glucose is present?

3. In bacterial operons such as the two discussed here, why is it not necessary that a gene that encodes a repressor protein be near the structural genes?

4. In the *trp* operon model, suppose that one of the encoded proteins had the ability to interact with the attenuation sequence and cause RNA polymerase to ignore the attenuator. Would this be an example of positive or negative regulation?

5. Is the attenuator region a positive or negative regulator?

6. What is the difference between an aporepressor and a repressor?

7. Besides coordinate regulation, what is another advantage of the operon model?

8. Suppose the gene that encodes adenylate cyclase (the enzyme that catalyzes the formation of cAMP) becomes mutated. What effect would this have on the regulation of expression of the lactose operon?

9. How does the binding of the *lac* repressor to the *lac* operator prevent transcription?

10. There are two stem-loop structures that can be formed by regions of the *trp* operon attenuator sequence. Why are the formation of these structures mutually exclusive?

11. Compare and contrast the activity of the *lac* operon repressor with that of the *trp* operon.

12. What would happen to the regulation of the *trp* operon if the two *trp* codons, in the leader region of the mRNA, were both mutated so that they no longer coded for *trp*?

13. Would it be possible to complement a mutation in the *trp* operon leader region that results in a defect in tryptophan biosynthesis control?

Answers to Chapter 11 Study Questions

1. A mutation in either the *lacI* gene or the operator would prevent binding of the repressor to the operator and would have this effect.

2. With no lactose and no glucose, both the repressor and the cAMP-CRP complex are bound. When only lactose is present, only the cAMP-CRP complex is bound. When only glucose is present, only the repressor is bound.

3. Repressor proteins are soluble and diffuse throughout the cell. Therefore they can act at a distance.

4. Positive regulation.

5. Negative regulator.

6. An aporepressor requires the presence of a corepressor in order to bind to the operator, whereas a repressor is functional on its own.

7. Another advantage of the operon model is the conservation of nucleotide sequences in DNA. In an operon, one set of regulatory sequences (e.g., promoter) serves more than one structural gene. For prokaryotes, the "efficiency" argument, presented in Chapter 1 of *EMB*, is often employed to explain various features of metabolic processes and is relevant here.

8. Assuming that the mutation resulted in complete loss of adenylate cyclase activity, no cAMP would be available for cAMP-CRP complex formation. The lack of this positive regulator would prevent transcription of the *lac* operon.

9. The *lac* repressor physically blocks the binding of RNA polymerase to the promoter.

10. The two structures are mutually exclusive because one region of base-pairing is common to both structures.

11. The repressor of the *lac* operon binds to the operator unless prevented from doing so by the inducer (lactose). Conversely, the repressor of the *trp* operon binds to the operator only when bound to the corepressor (tryptophan). Both repressors are necessary to inhibit the initiation of transcription.

12. Attenuation control would be lost, leaving only the repression system in regulating the *trp* operon.

13. Yes, but only if the *entire* operon were used to complement the original, defective operon.

Concept Map

To understand the "big picture," maybe what we need is a concept map. Let's prepare one!

Positive Mode

Negative Mode

Translational Control

Regulation During mRNA Synthesis

Posttranscriptional Control

lac Operon

Galatose Operon

mRNA Stability

trp Operon

Proteolysis

Feedback Inhibition

Here is a list of components that should be included in your concept map:

DNA looping
activator-inducer-repressor
constitutive mutant
cAMP-CRP

isozyme
attenuator
autoregulation
branched pathway
allosteric change

Practical Applications

Research in the regulation of prokaryotic gene activity has practical applications in at least two senses. First, in a general sense, research in the regulation of bacterial gene expression (also viral, see Chapter 12 of *EMB*) provides background information that can be utilized by genetic engineers. Recall, using recombinant DNA technology, various human genes (e.g., insulin) can be introduced into bacteria. With the proper regulatory sequences attached, the foreign gene can be expressed at such high levels that a commercial product can be harvested. The baseline information gained from the study of gene expression in bacteria has been extremely valuable for genetic engineers.

Second, in a more direct sense, knowledge of the regulation of prokaryotic gene expression has resulted in another practical application. Through an understanding of the general features of gene regulation described in Chapter 11, researchers have been able to achieve better yields of product in their fermentation tanks. Fermentation processes represent important ways to produce a wide variety of commercial products, ranging from drugs to foodstuffs. Microbial geneticists, using information about regulatory genes, have been able to optimize the yield from fermentation processes by employing novel strains of microorganisms. For example, production of the widely used antibiotic streptomycin, generated by the microorganism *Streptomyces*, has been greatly enhanced by selecting special strains, the most productive of which often exhibit minor changes in their regulation of several genes within the antibiotic synthesis pathway.

Solution for Exercise!:

I

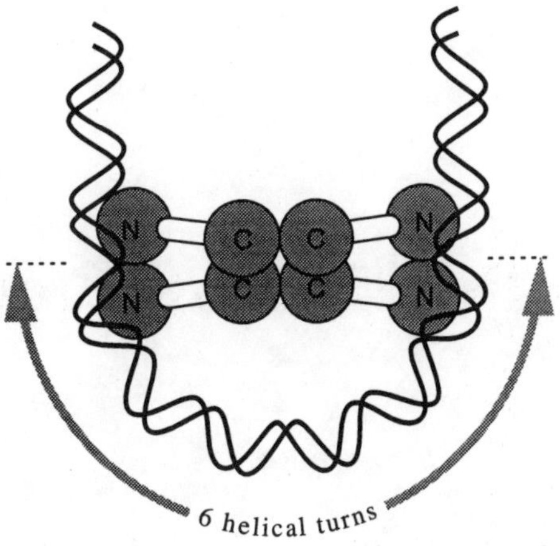

N
C
C
N
N
C
C
N
6 helical turns

II.

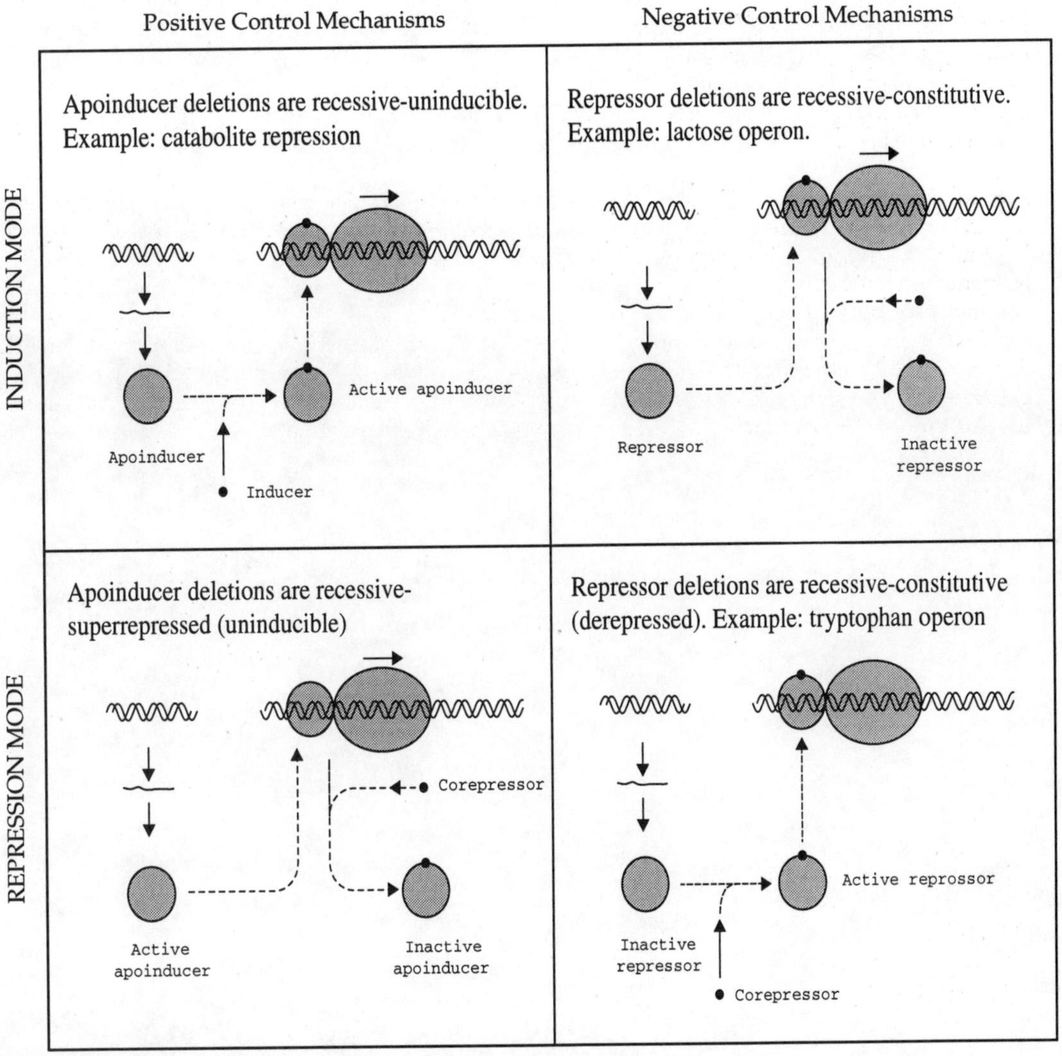

Positive Control Mechanisms
Negative Control Mechanisms
INDUCTION MODE
Apoinducer deletions are recessive-uninducible. Example: catabolite repression
Active apoinducer
Apoinducer
Inducer
Repressor deletions are recessive-constitutive. Example: lactose operon.
Repressor
Inactive repressor
REPRESSION MODE
Apoinducer deletions are recessive-superrepressed (uninducible)
Corepressor
Active apoinducer
Inactive apoinducer
Repressor deletions are recessive-constitutive (derepressed). Example: tryptophan operon
Active reprossor
Inactive repressor
Corepressor

12

Bacteriophage Life Cycles

Here's Help

Why does this chapter adopt the strategy of reviewing so many different phage types? Why not just focus on one?

"Diversity" is the key concept presented in this chapter. Viruses differ in size, morphology, and life cycle. No single diagram of a virus life cycle or of a mature virus particle can represent the complete spectrum of types. Just look at Figure 12-1 in Chapter 12 of *EMB* to gain an appreciation for phage diversity in morphology. Table 12-1, on the same page, further illustrates differences among various phages.

What sort of genes are generally carried on the viral genome?

First and foremost, the genes that code for the proteins that comprise the protective coat of the virus are carried on its genome. Each virus has its own coat proteins. Some coats are elaborate, consisting of a head, tail, and tail fibers (e.g., phage T4), while others are amazingly simple, consisting of multiple subunits of a single type of protein (e.g., phage M13).

Second, genes that code for the synthesis of components that are unique to the virus, not normally found in the host bacterium, are carried on its genome. For example, Phage T4 codes for enzymes that synthesize 5-hydroxymethylcytosine, as described in Chapter 12 of *EMB*.

Third, some of the more ornate viruses (e.g., T4) encode proteins that play a role in assembling the various coat proteins into a mature structure.

Finally, some viruses carry genes involved in either lysogeny or transduction. The number and forms of genes varies from one type of phage to another, as the data in Table 12-1 reveal.

M13 virus particles are excreted through the bacterium's cell membrane?

Yes, really. Instead of synthesizing an enzyme that lyses the bacterial host, thereby releasing the newly made viruses, M13 is excreted through the host's cell membrane. This process represents a clever evolutionary adaptation on the part of the virus. Newly synthesized viral DNA complexes with a specific DNA-binding protein. This complex then associates with the inner surface of the bacterium's cell membrane. There in the cell membrane, accumulated earlier during the viral infection cycle, reside coat proteins. The viral DNA assembles into the coat proteins. The completed phage particle then passes through the cell wall to the outside. Amazing, eh?

What determines whether a λ phage infection will result in a lysogenic or a lytic life cycle?

There is no short, simple answer to this question. We will, therefore, develop an answer by providing a stepwise description of both the lytic and the lysogenic pathways. Please refer to Chapter 15 of *EMB* for details.

The lytic cycle of λ phage:

1. A phage attaches to a bacterium and injects its DNA.

2. Once inside the cell, the DNA molecule assumes a circular form.

3. Three different transcripts are then produced:
   a. from pL to tL1, the protein product has been given the name N.
   b. from pR to tR1, the protein product is the repressor called "Cro."
   c. from pR' to tR3, no product—tR3 prevents RNA polymerase from reaching any structural genes.

4a. N protein binds to *nut* sites. Therefore:
   a. polymerases that start at pL no longer stop at tL1, so that in addition to the gene for N, the genes for recombination enzymes are transcribed.
   b. polymerases that begin at pR no longer stop at tR1, so that in addition to the gene for *cro*, the genes for O, P, Q, and *cII* are transcribed.

   (If growth conditions are unsuitable, cII will not be degraded and the phage will follow the lysogenic pathway).

4b. With increasing *cro* concentrations, *cro* begins binding not only to oR3, but also to oR1 and oR2. This decreases the amount of transcription from pR and, thus, lowers production of *cro*, O, P, and Q.

5. Q, the antiterminator produced in step 4a, binds to *qut*. This allows the polymerases that are beginning transcription at pR' to proceed through tR3 and transcribe the genes encoding proteins needed for cellular lysis, and for the heads and tails of progeny phage. Thus, the lytic cycle proceeds.

The lysogenic pathway of λ phage:

—assume steps 1,2,3, and 4a above have already occurred—

1. If *cII* isn't degraded, it activates transcription of the repressor *cI* (from the promoter pRE) and *int* (from pI). *Int* is necessary for integration of the phage DNA into the host DNA.

2. *cI* binds to oL, oR1, and oR2.
   a. this inhibits further transcription from pL and pR (i.e., stopping the lytic pathway at step 4a—preventing step 5 from occurring).
   b. the binding of *cI* to oR2 activates transcription at pRM, increasing production of *cI*.

3. When levels of *cI* are high, some *cI* is able to bind to oR3, and decrease transcription from pRM, which then decreases production of *cI*. It is this constant presence of a certain level of *cI* that perpetuates the lysogenic state.

Here's a brief summary, or inventory, of λ phage regulatory proteins.

N: acts as an antiterminator at tL1, tR1, and tR2.

Q: acts as an antiterminator at tR3, making possible the transcription of the middle and late genes.

cro: a repressor: low concentrations of *cro* repress pRM and high concentrations also repress pL and pR.

cI: at low concentrations, *cI* activates pRM and represses pR and pL, but at high concentrations, it represses all three. The degradation of this protein induces the lytic pathway.

cII: this protein activates pRE and pI; high concentrations favor the lysogenic pathway.

cIII: this protein helps maintain the lysogenic state by stabilizing *cII* and preventing its degradation.

Clearly, the regulatory circuitry associated with the λ phage life cycle is complex. One of the reasons for studying phage in such great detail is to gain insight into the nature and complexity of biological processes. In many instances, the circuits seem overly complex. Recall the "efficiency argument" discussed in Chapter 1 of *EMB* . When we learn more about phage life cycles, we will probably understand that they are indeed efficient and fine-tuned!

How important is the sequential expression of phage genes?

Generally speaking, phage genomes contain three classes of genes: those whose products are involved in integration (temperate phage only) and replication, those whose products are involved in structure, and those whose products are involved in cellular lysis. If we pause to consider the minimalistic (i.e., highly efficient and simple) aspects of the phage system, then the need for expression of the above types of genes in a fixed order becomes apparent. Remember, genes in "higher organisms" are usually reversibly regulated. That is, they can be "turned on or off" in response to the changing needs of the cell. Phage genes are, however, usually activated only once in the life of that phage. This is where sequential expression becomes necessary; it would not be particularly advantageous to the viral life cycle if the proteins responsible for cellular lysis were manufactured before the structural proteins that comprise the head of the phage.

Any genes encoding proteins involved in replication or integration of the phage genome are usually expressed early in the life cycle of the phage. Hence, the term "early genes" is used to categorize this class of genes. After expression of the early genes has reached a certain level, the "middle genes" begin to be expressed. These middle genes include genes that code for the protein structure of the phage. The last genes to be expressed, called "late genes," are those that encode proteins involved in host cell lysis.

How is this sequential expression achieved?

Although there are three basic mechanisms by which sequential expression of phage genes is accomplished, only two are utilized by the phages discussed in Chapter 12. Therefore, we will restrict our focus accordingly. No matter which mechanism is employed to insure sequential expression, there is one basic motif: Expression of the early genes produces some "factor" that allows or directs transcription of the middle genes. Likewise, the expression of the middle genes results in the presence of some factor that causes transcription of the late genes. The middle genes are transcribed only when that factor encoded by one of the early genes is present, and the late genes are transcribed only when that factor encoded by one of the middle genes is present. Thus, proteins needed for cellular lysis will not be present unless the middle genes have been expressed. The actual mechanisms for this ordered expression are quite simple and elegant.

Phage T7 exemplifies one of the basic mechanisms that insures ordered gene expression. The host RNA polymerase is employed to transcribe the early genes of the phage genome. One of these early genes encodes a phage RNA polymerase that only recognizes the promoters for the later genes. As the host RNA polymerase can only recognize the promoters of the early genes, the later genes will not be transcribed until the viral polymerase is present.

The second mechanism of insuring sequential expression that we will discuss here is used by both T4 and λ phage. These phages use the host RNA polymerase to transcribe all genes— early, middle, and late. By either using molecules that modify the polymerase in some way (such as an antiterminator like Q or N in the λ system), or by the presence of a phage-encoded promoter recognition subunit of RNA polymerase (the sigma subunit), ordered expression is insured. For instance, in the λ system, until the antiterminator N is expressed and binds near tL1, the genes that encode proteins involved in the integration of phage DNA into the host genome will not be transcribed. Another example, this one hypothetical, might help illustrate this mechanism. Recall, the sigma subunit of RNA polymerase enables recognition of certain promoters (see Chapter 8 of *EMB*). As might be expected, different sigma subunits confer specificity to different promoters. Suppose that only the early genes of the phage genome had promoters recognizable by the bacterial-encoded sigma subunit. Because of this, the early genes would immediately be transcribed by the host RNA polymerase. Also, suppose that one of the early genes encoded a sigma subunit that can complex with the host RNA polymerase and allow recognition of middle gene promoters. Likewise, suppose that one of the middle genes encoded a sigma subunit that confers specificity to promoters of the late genes. This last example is not meant to be representative of any particular phage, but rather to help you to better understand how this sort of mechanism might work!

How does a transducing phage differ from a "regular" phage?

When progeny phage are being produced, phage DNA (assuming that the phage has a DNA genome) is packaged into the phage head. Some types of phage, however, occasionally package host DNA instead of their own. A phage in which this can occur is called a transducing phage, and a phage head filled with bacterial DNA is likewise called a transducing particle. There are two types of transducing phages: "specialized transducing phage" and "general transducing phage."

Specialized transducing phage are so named because they can only package certain sequences of host DNA. All phages of this class, of which λ is an example, must be able to proceed through the lysogenic pathway. In fact, these phages must follow this pathway to produce transducing particles. Transducing particles are formed by aberrant excision of the prophage. That is, the host DNA adjacent to the integrated phage DNA is excised and packaged, instead of packaging the complete prophage itself.

General transducing phages, exemplified by phage P1 and phage P22, produce transducing particles that contain any region of the host DNA. The way in which these particles are produced is very straightforward. Before the packaging of DNA into the phage heads occurs, these phages cause the host genome to be partially digested, or fragmented. Since DNA is inserted into the phage heads based only on size, occasionally a small fragment of host genome, similar in size to the phage genome, gets packaged into a phage head.

Key Terms

Lytic cycle

Lysogenic cycle

Virulence

Temperate

Catalytic proteins

Structural proteins

Lysozyme

HMC

Terminal redundancy

Circular permutation

Headful mechanism

Temporal regulation

Antiterminators

Terminase

Prophage

Induction

Excisionase

Integrase

Transducing particles

Transducing phage

For More Information . . .

Additional information on these topics is indeed available:

Chapter 17 of Watson, *et. al., Molecular Biology of the Gene,* 4th edition (1987), explains phage in detail. Figure 17-15 of that chapter has an especially good diagram of the λ life cycle.

In addition, another helpful reference is pp. 273-287 of Alberts, *et. al. Molecular Biology of the Cell,* 3rd edition. Viruses and transposable elements are explained well here.

The Wise Owl Says

Simple experimental systems permit simple answers to simple questions. The simplest answers are, of course, "yes," "no," or a number.

Bacteriophage were first exploited as experimental systems by physicists and chemists who were intent on studying the simplest life forms. They reasoned that a complete understanding of the *principles* of life would emerge from analyses of organisms that were capable of performing only fundamental tasks, such as reproducing themselves.

Max Delbruck (1906-1981), a German born and educated scientist, is usually credited with establishing the "phage school" of biology. During the 1940s he performed a number of key experiments on bacteriophage reproduction. These experiments eventually (in 1969) led to his sharing of a Nobel Prize with Salvador Luria and Alfred Hershey, two other prominent members of the early "phage school."

This "simple system" strategy clearly worked. A phage was the first organism for which the complete genome was mapped. From that success emerged the expectation that it is possible to obtain a complete set of genetic information from any organism. Therein lies the rationale for the gigantic human genome sequencing project, now nearly one-third complete.

Exercise

I. Let's construct a "minimalist" virus in a stepwise fashion. That is, the smallest, simplest virus that we can possibly conceive.

1. Review Table 12-1 to find the virus with the smallest genome.

2. Beginning with that size, calculate how many nucleotides the viral genome contains.

3. Calculate the number of nucleotides required to code for a single coat protein:

    a. How many amino acids should a small protein contain?

    b. From (a.), estimate the number of nucleotides required.

4. Predict which, if any, enzymes the virus might need for its replication. Consider the following possibility. The smallest viruses (Table 12-1) have an ssRNA genome. Wouldn't an RNA replicase enzyme be required? Isn't it unlikely that the host bacterium would have one, since it normally has no need to replicate ssRNA?

5. Estimate the size of an RNA replicase molecule. Keep in mind that it should not be too small, since RNA replication is likely to be complex. Perhaps a small-medium size protein would do. How large would that be?

6. Consider whether any additional proteins would be required for viral replication that the host bacterium would not provide. Calculate the number of nucleotides required to code for those additional proteins. Recall, however, this is an extraordinarily simple virus, so no additional replication proteins may be required.

7. Now, total up the number of required nucleotides. How does your estimate compare with the known size of the MS2, Qß, and f2 genomes (Table 12-1).

If your calculations yield a genome size smaller than 1×10^6, suggest an experimental protocol that could be employed for undertaking a search for this "minimalist" phage.

II. Write a brief essay on the bacteriophage life cycle, in which you combine information from at least three *EMB* chapters. For example, weave into your essay the information included in Figure 5-13 in Chapter 5; Figure 7-9 in Chapter 7; and several sections of Chapter 12. Please try to develop the "diversity" theme featured in Chapter 12.

Far Out

Why don't we use information of the type presented in Chapter 12 of EMB to design mammalian viruses for human gene therapy? Recall, retroviruses are potentially useful for this purpose. Perhaps it would be possible to extend their usefulness by custom-designing them to infect only a particular tissue or organ. This way patients could be injected with a virus vector that carries the desired "correcting gene." This has already been accomplished with cultured cells. For example, a mammalian (e.g., human) cell line derived from lung tissue has been infected with the retrovirus that contains a complete copy of the cystic fibrosis transmembrane conductance regulator gene. Some of the cells integrated this gene into their genome and displayed a diminution in the cystic fibrosis phenotype.

Now, the challenge is to design viruses capable of infecting only a specific tissue or organ. This way it might be possible to deliver correct copies of disease-causing genes and thereby rescue many patients. Wouldn't you expect muscle diseases (e.g., Duchenne muscular dystrophy) and metabolic diseases (e.g., enzyme deficiencies) to be amenable to this sort of treatment?

So, where can we begin designing this virus? The key step is the adsorption of the virus to the target (diseased) tissue. Chapter 12 of *EMB* mentions that phage adsorb to specific receptors on the bacterial surface. Maybe an analysis of the viral adsorption process would be a good place to begin. Since adsorption may occur by fusion of the viral protein coat with the cell membrane, perhaps the complementarity between these two membranes would be a good experimental starting point for developing a tissue-specific infecting virus. From there on it would probably be a matter of learning how to engineer the virus so that the rescue gene would be stably expressed.

Possible? For sure. Far out? No doubt!

Study Questions

1. What role does lysozyme play in the viral infection cycle?

2. What protein would not be made in a λ Q^- mutant?

3. Why do phage usually require specific receptor molecules on the surfaces of bacteria for adsorption?

4. Since infection by phage usually kills the cell, why haven't bacteria that lack anchorage molecules evolved?

5. What role does the enzyme DNA ligase play in the initial stages of infection of a cell by λ phage?

6. In a λ N^- mutant, would any *cro* protein be synthesized? *cII* protein?

7. What role does HMC play in the life cycle of T4?

8. What purpose does the enzyme dCTPase serve in the life cycle of phage T4?

9. What is the difference between T4 and T7 in the way they maintain a sequential expression of phage genes?

10. What would happen if all the genes of a virulent phage (such as T7 or T4, for example) were expressed simultaneously?

11. Why wouldn't a cell with a λ prophage be lysed as a result of infection by a second λ phage?

Answers to Chapter 12 Study Questions

1. Lysozyme is synthesized late in the replication cycle and digests the bacterial cell wall, releasing newly synthesized phage particles.

2. Since transcription beginning at pR' would terminate at tR3 when Q isn't present, no proteins needed for the heads and tails of progeny phage would be made, nor would any of the proteins needed for cellular lysis.

3. In order to physically stabilize the phage and enable it to inject its DNA, the phage needs an anchor. The receptor molecules provide these anchorage points.

4. Because the molecules to which phage attach are usually proteins that are either receptors or transporters of some substance (e.g., glucose, maltose) necessary for bacterial growth and survival. These molecules just happen also to be points of attachment for certain phage.

5. It seals the nicks in the backbone of the circularized form of the λ DNA (formed by the overlapping and base-pairing of the cohesive ends of that molecule).

6. Cro protein would be made, but since termination would occur at tR1 without the presence of N, the *cII* gene wouldn't be expressed.

7. It serves to distinguish phage DNA from host DNA. The phage synthesizes nucleases that degrade DNA containing naked cytosine. Thus, host DNA is degraded, while phage DNA is protected from the nuclease activity since its cytosines are methylated.

8. It insures that cytosine is not incorporated into the DNA of the phage.

9. T7 initially employs the host RNA polymerase to transcribe the early genes; a phage-encoded RNA polymerase is then used for the later genes while that of the host is inactivated. T4 uses the host RNA polymerase throughout, but modifies it repeatedly.

10. Lysis of the cell would occur before many (if any) progeny phage are assembled.

11. The repressor that maintains the prophage in the lysogenic state would be present, preventing the second phage from entering the lytic pathway.

Concept Map

How do the phage biographies of Chapter 12 fit together? Let's make a concept map and find out!

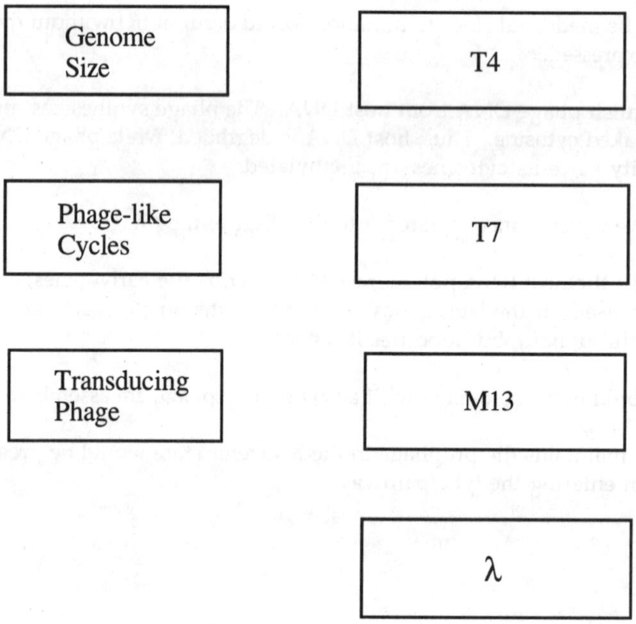

Please include the following components in your map. You may also want to scan the "Key Terms" list for additional components:

- immunity
- virulent
- HMC
- lysogenic life cycle
- ds, lin-ss, circ-ss, lin-ds, circ
- Cro protein
- prophage
- headful mechanism
- temperate
- class I, II, III transcripts
- RFI

Practical Applications

Many of the viruses described in Chapter 12 of *EMB* serve a double duty for molecular biologists. First, they provide simple model systems for studying gene regulation. Several of the principles discovered in these simple phage are expected to help us understand the control of gene expression in eukaryotic organisms.

Second, many of the viruses included in Chapter 12 are being exploited as vectors for genetic engineering experiments. The phage genome can often accommodate relatively large DNA inserts, compared to the accommodations of plasmids. For preparing genomic DNA libraries, it is important to clone large fragments. Hence, several viruses, especially lambda, have been useful for this purpose.

Other phages are useful vectors for procedures such as DNA sequencing. M13 phage (see Figure 12-12 of *EMB*) has a single-stranded genome into which the test DNA can be cloned and later removed. Since the phage DNA is single-stranded, it can be sequenced directly using the Sanger procedure described in Figure 3-18 *EMB* .

13

Regulation of Gene Activity in Eukaryotes

Here's Help

What is the purpose of this chapter?

As previously stated, most of *EMB*'s information has been derived from the study of prokaryotic molecular biology. In this chapter, however, our attention is turned to eukaryotes. The first topic to be addressed is "control of gene expression," which is often quite different in eukaryotes. Prokaryotes are unicellular organisms that grow and divide rapidly. Almost all of a prokaryote's genes are expressed for the majority of the cell's life. In multicellular eukaryotes, however, cells are often very specialized and live longer than do most prokaryotic cells. Because of this specialization, most cells of a multicellular eukaryote express only about 7% of their genes. The genes that are not absolutely necessary to that particular cell are not expressed (for example, a neuron would not normally express the gene for one of the hemoglobin subunits). Areas of the genome that are not expressed in a particular cell (of course, this varies from cell type to cell type) are maintained in a compact state and are collectively referred to as heterochromatin. Those areas of the genome that are actively expressed are maintained in a less condensed form, allowing the enzymes and other proteins necessary for expression access to the genes. These regions are referred to as euchromatin.

There are two distinct characteristics of euchromatin: sensitivity to the enzyme DNase I and hypomethylation. These characteristics can be used by molecular biologists to distinguish between euchromatin and heterochromatin. DNase I is an enzyme that digests free, or naked, DNA. Since heterochromatin is highly compact and associated with many proteins, it is not digested by this enzyme. Euchromatin is much more accessible and, therefore, will be degraded. The second characteristic of euchromatin is hypomethylation. Eukaryotic DNA is usually methylated on cytosine residues that are followed by guanine residues (i.e., 5'-CG-3'). This is clearly illustrated in Figure 13-3 of *EMB*. However, euchromatin is less methylated than heterochromatin. The full effect of methylation is not known, but it has been shown that genes with methylated promoters cannot be transcribed.

Three eukaryotic RNA polymerases?

Eukaryotes, unlike prokaryotes, have three different types of RNA polymerases. These three enzymes are known as RNA pol I, II, and III. They were mentioned previously in Chapter 8. RNA pol I transcribes RNA that will be cleaved into rRNA, pol III transcribes the tRNA genes, and RNA pol II transcribes everything else (i.e., all genes that code for proteins). Hereafter, we will limit our discussion to pol II.

Pol II is different in many ways from the *E. coli* RNA polymerase that we studied in Chapter 8 of *EMB*. The prokaryotic polymerase is able to recognize DNA promoter sequences, enabling it to "know" where transcription of the DNA molecule should begin. Eukaryotic pol II, however, cannot recognize the promoter sequences unless certain DNA-binding protein molecules, called "transcription factors", are associated with them.

How is regulation of transcription stimulated?

One way that gene expression can be regulated is by control at the level of transcription. There are basically two types of "elements" that control the transcriptional activity of pol II. First, there are special DNA sequences, called promoter and enhancer sequences, that act as regulators. Second, there are protein molecules, called "transcription factors", that interact with promoters, enhancers, and pol II itself to help activate transcription. As you should recall from Chapter 8 in *EMB*, promoters are DNA sequences near the point of initiation of transcription. The eukaryotic promoter consists of a TATA box, located at about -29, a

CAAT box, and GC-rich sequences that are even further upstream. Pol II cannot recognize "naked" promoters. Specific transcription factors (proteins) bind near various parts of the promoter. It is believed that this promoter-transcription factor complex guides pol II to attach and begin transcription of the downstream gene. (Remember: In prokaryotes, the polymerase "simply" recognizes the promoter and binds to it; in eukaryotes, the polymerase only recognizes promoters that have transcription factors bound to them.)

Promoters are absolutely necessary for transcription and must be near the starting point of the gene to be transcribed. Enhancers, on the other hand, are DNA sequences that "strengthen" the transcription of a gene. Together with transcription factors, enhancers interact with polymerase at the gene's promoter sequence and thus increase the frequency of transcription. One difference between enhancers and promoters is that enhancers can be located comparatively far away from the gene whose transcription they affect. As illustrated in Figure 13-2 of *EMB*, enhancers located either upstream or downstream can be effective. How? Probably the DNA molecule forms a loop, bringing the enhancer closer to the promoter. Another difference between enhancers and promoters is that enhancers are not absolutely required in order for transcription to occur, they just increase the frequency at which a gene is transcribed.

In a nutshell, how is transcription initiated?

Pol II goes up and down a DNA molecule until it finds a promoter associated with transcription factors. It then binds to the promoter to begin transcription, and is given an extra "push" by one or more transcription factors that are bound to enhancer sequences.

What is RNA processing?

Eukaryotic genes usually consist of both exons, that encode the gene products, and introns, that do not appear in the final gene product. The whole gene, including both introns and exons, is transcribed into RNA. In processing, the exons are linked, while the intervening introns are "cut out." Such processing does not exist in prokaryotes and the discovery of introns came as a big surprise to molecular biologists.

What is gene splicing?

Most eukaryotic genes contain introns that interrupt the gene coding sequences. The mRNAs that result from direct transcription need to be processed, to remove the intron sequences, before they can be translated.

"Splicing" is the term given to this process of cutting out the intron sequences from the primary transcript. The cutting machinery is likewise called a "spliceosome." Specific nucleotide sequences, so-called splice sites, are recognized by the cutting-ligation enzymes of the spliceosome. These are described in Chapter 8 of *EMB* and in Figure 13-4 of *EMB* . Once spliced, capped, and tailed (see Chapter 8 of *EMB*), the primary transcript is ready for translation.

What then is *alternative* splicing?

Figure 13-5 of *EMB* illustrates four different ways that a primary transcript can be cut and ligated. The text explains in detail these various splicing patterns.

Please keep in mind the advantage of alternative splicing to a cell. Several slightly different proteins can be produced from one large primary transcript. Therefore, the versatility of the genome is expanded by virtue of the possibilities for alternative splicing.

How about some help in understanding Figure 13-5?

For sure! The angled lines above or below each strip of mRNA primary transcript indicate the splicing pattern. For example, in (a) the region from the right edge of the first ("left-most") exon and the third exon is cut out by a spliceosome (and discarded). Spliceosome action is shown in Figure 13-4 of *EMB*. As well, on the same primary transcript, in the same processing scenario (top set of angled lines), the intron region between that third exon and the fourth ("right-most") exon is cut out by spliceosome action and discarded. Thus, the processed mRNA shown in the right (lower) diagram would have been produced.

What is the "take-home" message from Figure 13-6?

The ability of some mRNA's to function is regulated, in various ways. Figure 13-6 shows four specific examples of the types of ways employed by the cell to control the extent to which an mRNA is translated into protein. No doubt countless other ways also exist, and await discovery.

How do glucocorticoids regulate gene expression?

Glucocorticoids control transcription by binding to receptor proteins. As illustrated in Figure 13-7 of *EMB*, the receptor-steroid complex in turn binds to an enhancer sequence in the genome. By binding to an enhancer, the promoter region is activated and transcription of the nucleotide sequence adjacent to the promoter is initiated.

How does genome rearrangement increase antibody diversity?

The IgG molecule consists of two heavy (i.e., longer) chains and two light (i.e., shorter) chains. Each of these chains consists of two regions: a constant region, conserved in all IgG molecules, and a variable region, different in all IgG molecules. (Please return to Chapter 4 in *EMB* (Figures 4-9, 4-10 and 4-11) if you would like a brief review of immunoglobulin structure.) It is the variable region that confers antigen-specificity. The human body is capable of generating about 10^8 different antibodies. If our B lymphocyte cells adhered to the "rule" that each different protein must be encoded by a different gene, the cell would have to devote 10^8 genes to making antibodies alone. This is clearly impossible.

Cells have evolved a unique way of solving this problem. During each B lymphocyte's course of maturation, its DNA is rearranged and some segments are discarded so that it is only able to make one type of IgG molecule. That is, all of the IgG molecules produced are specific to one, and only one, antigen. How does this happen? Well, each light chain is encoded by a gene that is composed of three collinear regions: the V region, the J region, and the C region (these aren't separate genes, just different regions of one gene). The V and J regions together encode the variable segment of the light chain, while the C region encodes the constant segment. Immature B lymphocytes contain sequence for about 300 different V regions, 4 different J regions, and 1 C region. Different combinations of these regions would allow the cells to make light chains for every possible different type of IgG. Nevertheless, remember that each mature B cell makes only one type of IgG.

This is where genome rearrangement comes into play. Instead of just expressing the particular V region and the particular J region needed, and ignoring all other regions, the cell goes one step further. It actually "cuts out and throws away" those regions that aren't needed. It does this by a process of recombination. Suppose that a B cell has to make an IgG that will recognize a certain protein in the cell wall of a bacterium. To do this, the V region of the IgG light chain must be encoded by the 22nd V region and the 4th J region. Recombination allows the cell to "move" V_{22} next to J_4 by excising all intervening DNA. The VJ segment needed to encode the proper light chain variable region is now formed. To form a gene that will encode the entire molecule, recombination occurs between the end of the VJ segment and the beginning of the C region. The end result is a gene with a VJC coding sequence that encodes the proper IgG light chain. A similar process occurs in the formation of the heavy chain gene.

Key Terms

Chromatin

Monocistronic RNA

Intron

Exon

Promoter

Enhancer

Transcription factor

Hypersensitive site

Methylation

Heterochromatin

CpG island

Splicing

Spliceosome

Intron retaining mode

Exon cassette mode

Small nuclear ribonucleoprotein complex (snRNP)

Trans-splicing

RNA editing

Endoplasmic reticulum

Nuclear pore complexes

hnRNPs

Negative translational control

Translational frameshifting

Translocation signal

SRP

Ubiquitin

Glucocorticoid receptor

Immunoglobulin

Antigen

IgG

Allelic exclusion

Clonal section

Affinity maturation

For More Information . . .

Where can I read more on these topics?

Portions of Chapter 9 (pp. 401-404 and 417-419) in Alberts *et. al.*, *The Molecular Biology of the Cell,* 3rd edition, describe a variety of mechanisms that control gene expression. The generation of antibody diversity is explained on pp. 1221-1227.

Watson *et. al.*, *Molecular Biology of the Gene*, 4th edition, contains a description of eukaryotic gene regulation mechanisms on pp. 708-722. A full-page diagram of RNA processing is included on p. 720.

The Wise Owl Says

Nothing in molecular biology makes sense outside an evolutionary context.

Examples of molecular phenomena that are best understood in terms of "phylogenetic history" (i.e., evolution) include:

DNA as the genetic maternal (Chapter 6)
histone structure (Chapter 5)
introns-exons (Chapter 9)
antibody genes (Chapter 13)
plasmids and transposons (Chapter 14)
mutation rates (Chapter 6)
DNA repair (Chapter 10)
Triplet code (Chapter 9)

The list goes on and on and on...

Closely related organisms seldom contain completely different gene products. Usually their gene products differ only slightly from one another. These slight differences reflect the inherent mutation rates of the genome and the time (often measured in millions of years) over which these mutation(s) that generated phenotypic differences occurred.

Comparative analyses are, therefore, very useful. Instead of performing experiments on human genetics, which is neither legally, socially, nor ethically allowed, the genetics of small mammals, such as mice, are studied intensively. Then, by using the phylogenetic logic discussed in Chapter 1 of *EMB* , substantial insights into human molecular mechanisms can be gained.

Exercise

I. Using two separate primary transcripts, diagram the possibilities for splicing, both within each transcript (cis), and between the two transcripts (trans). *EMB* explains cis versus trans splicing:

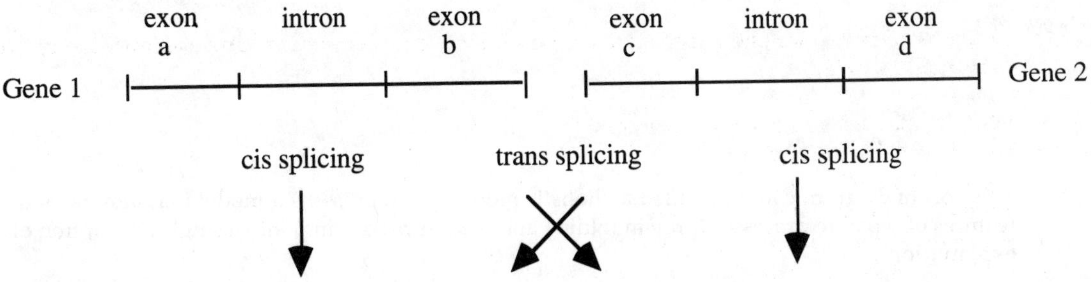

II. Illustrate the splicing reaction for two different primary transcripts, showing how the various snRNPs interact with the splice sites to form a spliceosome. That is, adapt the illustration in Figure 13-4 of *EMB* to show a splicing event that involves trans splicing as shown in Exercise I above:

III. In many instances, exons have been known to correspond with protein-folding domains. That is, an exon often codes for a segment of a polypeptide chain that folds into a particular configuration, thereby contributing to the functional properties of the protein.

Diagram a folding structure for a protein, based upon one of the examples shown in Figure 4-7 in *EMB*. Indicate on your diagram segments of the protein that might represent translational products from single exons. Now draw out a hypothetical gene structure that includes the predicted number of exons and introns located in the appropriate positions:

This sort of diagram could be called a "holistic model." It represents a model that integrates the features of separate processes (protein folding and mRNA processing) into a single description or explanation.

IV. Draw a diagram that illustrates the flow of a steroid molecule from outside a cell to its target, following logistical considerations:

How does the steroid enter the target cell? Do all steroids enter all cells?

Do all cells, or just target cells, contain free receptor proteins?

Do only receptor molecules that have bound steroids enter the cell's nucleus?

Once inside the nucleus, how does the steroid-receptor complex manage to locate the appropriate enhancer sequence?

Tough Nuts

I. An immune response involves interactions between many different types of cells. As mentioned in Chapter 13 of *EMB*, an antibody-producing cell is called a B-lymphocyte. The production of an antibody is a complex process that involves more than gene rearrangement alone. It involves interactions and transfer of information between several different types of cells.

Initially, one type of cell, usually a scavenger cell called a macrophage, engulfs the antigen and processes it. That is, the antigen is usually cut into fragments, which are then transported to the macrophage surface, where they are presented to other cells in the immune system. Next a T-helper cell takes over. It recognizes the antigen fragments presented on the surface of the macrophage, and secretes special chemicals that stimulate both the B cells and T cells to proliferate. Finally, the B cells produce and secrete antibody.

A tough nut can be found in virtually each step of antibody production. Each step is complex and poorly understood. Why? At least two explanations account for our lack of understanding:

1. Antigen diversity is enormous. No single processing mechanism could possibly handle all types of naturally occurring antigens.

2. Inherently complex gene rearrangements occur. To accommodate the diverse antigens found in nature, gene rearrangement has evolved as the molecular mechanism. This process is described in Figures 13-8 and 13-9 respectively in Chapter 13 of *EMB*. The molecular recognition processes and the information transfer steps that lead to antibody synthesis from initial antigen processing are poorly understood.

First, molecular biologists now accept that the production of an antibody probably represents a culmination of virtually all metabolic processes that a cell is capable of carrying out (e.g., division, interaction, movement, and specialization). Therefore, each step needs to be studied intensively. This approach is now in progress, and related topics, such as the T-cell response, are being researched from a component-processes point of view, such as antigen receptor synthesis information transfer to a B cell. Understanding each process represents a challenge.

Second, molecular biologists are attempting to develop a holistic (i.e., "big picture") view of how the immune system works. Their aim is to take all of the component processes and integrate them into a single whole.

Third, attempts are being made to understand the molecular basis of various immune-system-related diseases. Of special interest are autoimmune diseases that cause the immune system to recognize normal body components as foreign.

Fully understanding the immune system represents one of molecular biology's greatest challenges!

II. Cancer is one of the toughest nuts in eukaryotic gene regulation that molecular biologists have yet encountered. In simple terms, cancer represents the loss of normal growth control by a cell or tissue. Molecular biologists often refer to the change from a normal growth phenotype to an "out-of-control" growth mode as "transformation." In some cases, transformation occurs spontaneously, while in other instances it is caused by chemicals (called "carcinogens") or (in a few instances) by infection with a tumor virus. In all instances, however, this transformation is permanent. Cancerous cells produce cancerous daughter cells when they divide. It is therefore believed that the molecular basis of transformation involves changes in DNA, more specifically, alterations in nucleotide sequence.

Studies on tumor viruses led to the discovery of "oncogenes," genes present in the viral genome that are responsible for transformation. Molecular biologists were astonished to discover that such oncogenes exist in normal cellular DNA. That is, human DNA contains an array of such oncogenes, perhaps several dozen (or more). Once activated, either spontaneously or by a chemical such as an environmental mutagen (e.g., ultraviolet light or hazardous waste), the oncogene is expressed and converts a slow growing cell into a rapidly dividing cell. As a result, a tumor develops.

A wide variety of proteins are encoded by oncogenes, including proteins that directly stimulate cells to divide (so-called "growth factors"); proteins that act as receptors for growth factors (in the same way the glucocorticoid receptor in Figure 13-7 of *EMB* acts); and various DNA-binding proteins of yet unknown function.

Hence, several tough nuts arise:

1. How do individual oncogenes act to alter cellular metabolism, thereby causing cancer? Does each oncogene act differently?

2. What is the molecular basis of cancer susceptibility? Why do some cancers tend to run in families?

3. How many different oncogenes exist? A few dozen, or several hundred?

4. The toughest nut of all, of course, concerns reversing, or blocking, the effects of oncogene action. That is: HOW TO CURE CANCER?

Propose a way to crack these tough nuts!

Study Questions

1. Which of the following genes would most likely be expressed in all cell types and which would be restricted to specific cell types?
   a. a gene involved in glucose metabolism
   b. a gene coding for trypsin (digestive enzyme)
   c. a gene involved in myelin production
   d. a structural gene coding for a cell membrane integral protein
   e. a histone encoding gene

2. Why might transcriptional regulation be considered the most economical means of controlling gene expression?

3. What experimental evidence exists in support of the functional similarity between promoters and enhancers?

4. How does an enhancer exert its effect on the promoter if it is located thousands of base pairs away?

5. How does protein phosphorylation alter enzyme activity?

6. Which of the four different modes of alternative mRNA splicing involves modification of splicing activity? How do those modes that *do not* involve modification of splicing activity yield alternatively spliced mRNAs?

7. How are most intron-containing mRNAs retained in the nucleus? Do these mRNAs ever enter the cytoplasm? If so, how?

8. List at least three ways that a mature mRNA molecule can give rise to multiple protein products.

9. A certain highly stable mRNA molecule has a half-life of 2 hours and is maintained at a concentration of 1×10^{-4}M in the cytoplasm. Assuming that only the stability of the molecule changed, what must the new half-life be in order to maintain a concentration of 1×10^{-6}M?

10. Why is methylation suspected to be important in gene inactivation?

Answers to Chapter 13 Study Questions

1. All cell types: (a.) (d.) (e.).
   Specific cell types: (b.) (c.).

2. Transcription is the earliest process in the sequence of events leading from gene to enzyme activity. Hence, regulation at the level of transcription eliminates the possibility of overproduction by earlier processes, which would waste a lot of energy.

3. Experimentally multimerized promoters have been shown to behave as strong enhancers.

4. According to one model, the intervening DNA "loops out," bringing the transcriptional factors of the promoter and enhancer into close proximity.

5. The phosphate group has a strong negative charge that interacts with other charged groups on the protein. This interaction can significantly alter tertiary structure and, therefore, enzymatic activity.

6. The intron retaining mode and the exon cassette mode both involve modification of splicing activity. The differential promoter selection and differential cleavage-polyadenylation site selection modes both depend upon transcriptional factors for determining the splicing pattern.

7. Immediately after mRNA synthesis, splicing factors attach. This spliceosome assembly inhibits transport of the mRNA across the nuclear membrane. Intron-containing messages occasionally enter the cytoplasm via an unknown mechanism, possibly related to interference with spliceosome assembly, or by transport to a compartment free of splicing factors.

8. Translational frameshifting, posttranscriptional modification, and occasional skipping of the stop codon (e.g., Moloney leukemia virus).

9. 1.2 minutes (120 min./ 100)

10. Methylation is highly correlated with nontranscribed DNA. Also, experimental methylation of promoter regions has been shown to abolish gene activity.

Concept Map

An important concept map! Let's prepare a good one!

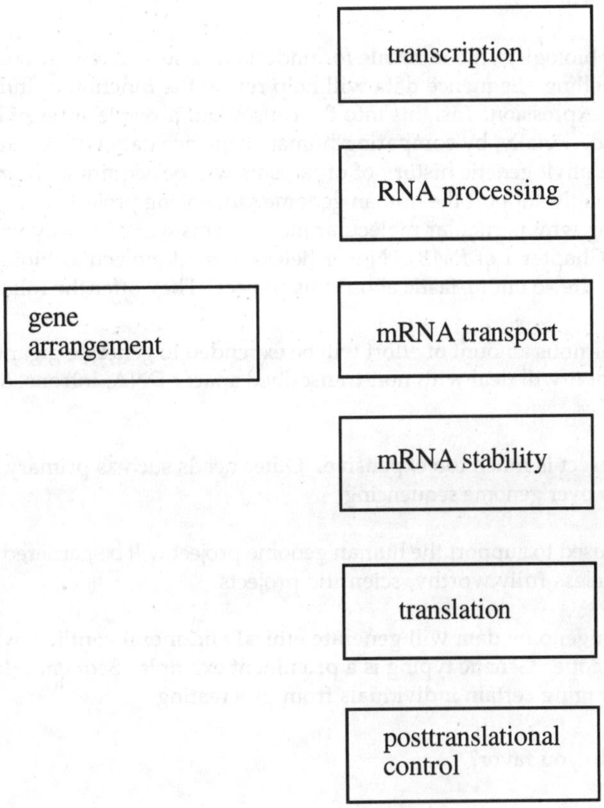

Please weave the following terms into your concept diagram:

RNA polymerase I, II, III
promoter
hnRNP
heterochromatin
enhancer
transcription factor

exon
nuclear pore complexes
translational
frameshifting
intron
hypomethylation
clonal selection
splicing

endoplasmic reticulum
trans-splicing
ubiquitin
half-life
negative control
immunoglobulins
allelic exclusion

Science and Society Issues

The human genome sequencing project, an effort designed to map and establish the nucleotide sequence of the entire human genome, is well underway. In fact, the project should be completed by the end of the present decade.

To molecular biologists, the rationale for undertaking such an enormous and expensive project is indeed compelling. Sequence data will help reveal the function of individual genes, as well as their regulation of expression. Insights into the causes and possible cures of human genetic diseases will also be gained. Finally, by comparing human sequence data with comparable data from other species, clues into the phylogenetic history of organisms will be acquired. Therefore, most molecular biologists strongly support the human genome sequencing project. Remember, these clues often aid in understanding why particular molecular mechanisms work the way they work (see the "Logic of Biology" in Chapter 1 of *EMB*). Nevertheless, not all molecular biologists, and certainly not all nonscientists, are so enthusiastic about this project. They offer the following counterarguments:

1. An enormous amount of effort will be expended to generate gigantic amounts of useless data. Much of it will deal with non-transcribed spacer DNA, introns, and other so-called "junk DNA."

2. The project is much too expensive. Other needs such as primary health care should receive priority over genome sequencing.

3. Funds used to support the human genome project will be garnered from other less prestigious, but nonetheless fully worthy, scientific projects.

4. Human genome data will generate ethical and moral conflicts with which society is not yet able to cope. Genetic typing is a prominent example. Sequence data could provide a basis for discouraging certain individuals from procreating.

Which side do you favor?

Practical Applications

As you may recall, the analysis of gene expression in prokaryotic organisms (Chapter 11 in *EMB*) generated basic research information that had practical applications. Those applications included various industrial fermentation processes, as well as biotechnology routines.

A similar situation exists for the study of eukaryotic gene regulation mechanisms. The basic information derived from these studies (described in Chapter 13 of *EMB*) can be applied to the development of molecular medicine procedures.

The techniques being developed for gene therapy depend heavily on basic information in eukaryotic gene structure and gene regulation mechanisms. Likewise, the design of cancer therapy strategies requires substantial amounts of basic information on gene regulation mechanisms associated with the control of cell proliferation. Finally, developing a cure for such neurological diseases as Alzheimer's Disease requires enormous amounts of background information on brain function.

Chapter 13 of *EMB* provides evidence that the requisite basic information on eukaryotic gene regulation mechanisms is being accumulated. Keep in mind, however, that the amount of detailed information available for eukaryotic regulatory mechanisms is small in comparison to prokaryotic organisms. Compare Chapter 13 of *EMB* with Chapters 12 and 11. As you can tell, much more experimentation on eukaryotic organisms is needed to achieve comparable levels of detail for eukaryotes!

14 Plasmids and Transposons

Here's Help

What's a plasmid?

A plasmid is a small piece of circular DNA that replicates autonomously within specific bacterial hosts. Plasmids encode no functions essential to the bacterial host, so plasmid-free bacteria live perfectly normal lives. In fact, the plasmid uses the host's enzymes, such as DNA polymerase, for its own replication.

What advantages do plasmids offer their bacterial hosts?

Plasmids, although not **necessary** for survival of the bacterial host, usually encode specific functions in their DNA that can benefit the bacterial host. R plasmids, for example, encode genes that make the bacterial host resistant to antibiotics, such as streptomycin and tetracycline. Also, F plasmids encode genes that confer the ability to form conjugation bridges to the bacterial host, enabling the host to transfer chromosomal DNA to female recipients (e.g., Figure 14-4 of *EMB*).

What is the difference between F$^+$, F$'$, and Hfr?

All refer to some aspect of the F plasmid's "life cycle." F$^+$ is used to indicate a cell that contains an F plasmid. This type of cell is also known as a "male" cell. Chapter 14 mentions that F plasmids can become integrated into the genome of the host cell by recombination. When this occurs, the cell is thereafter known as an Hfr cell. Occasionally, an F plasmid that has integrated into a host's cellular DNA will leave that DNA and resume its original plasmid form. When this occurs, some of the host DNA is often included in the plasmid. An F plasmid that contains host DNA is referred to as an F$'$ plasmid.

Are the genes carried on the F plasmid still active in an Hfr cell?

Most certainly! The F plasmid DNA doesn't know (or probably even care) whether it exists in a plasmid or Hfr form. However, problems arise for an Hfr cell when the F DNA, in the host chromosome, attempts to transfer itself via conjugation to a female cell. This F DNA ends up transferring not only part of itself into the recipient cell, but also some of the host DNA to which it is attached. Nevertheless, this has been of great use to molecular biologists. For example, the distance between two genes on a bacterial chromosome of an Hfr cell can be determined by allowing that cell to conjugate with another cell and measuring the time that elapses between the transfer of the two genes. Figure 14-5 of *EMB* illustrates the time course of gene transfer.

What's the difference between Figures 14-3 and 14-4 of *EMB*?

Figure 14-3 illustrates the transfer of *F plasmid DNA*, whereas Figure 14-4 represents the transfer of bacterial chromosomal DNA (that contains an integrated F plasmid) from the donor to the recipient cell. In Figure 14-3, the chromosomal DNA is not shown in order to simplify the illustration. That is why the F- female cell in the top three illustrations looks "empty." In Figure 14-4, however, the plasmid has already been integrated into the chromosomal DNA, so the chromosomal DNA is essentially all that is illustrated.

Simply stated: one illustration (Figure 14-3) shows plasmid transfer, while the other (Figure 14-4) shows chromosomal DNA transfer. Got it?

Figure 14-6 looks intimidating. Is it really that complex?

No, not if you remember the details of DNA replication that were presented in Chapter 7 of *EMB*. The upper series of diagrams illustrates typical unidirectional DNA synthesis. Replication begins on the diagram as a D loop (see Figure 7-6 of *EMB*) to the left of the origin of replication. Both a leading strand and a lagging strand are synthesized. One of the nice features of these illustrations is the labeling of all free 3'OH and 5'P ends of newly synthesized strands. Eventually, on the right-hand side of the upper set of illustrations, the whole circle is synthesized in *one* direction and the parental and daughter strands are released.

The lower series of diagrams is essentially the same as the upper series, except that DNA replication proceeds in both directions, hence the name "bidirectional." The lagging strand of the left-most replication fork becomes the leading strand of the right-most fork. This was illustrated earlier in Chapter 7 (please review Figure 7-24 of *EMB*). Eventually, as was the case for the upper illustrations, the parental and daughter strands are separated. The main difference? Bidirectional replication is much faster than unidirectional replication, so it is employed by plasmids which employ high copy numbers

What is the difference between an insertion sequence (IS) element and a transposon?

Both an IS element and a transposon have the ability to move from one location in a plasmid or chromosome to another location. However, IS elements are simpler in structure than transposons. IS elements consist of the gene(s) that code for the transposase enzyme(s), flanked by an inverted repeat. That is, either end of an IS element contains a repeated, and inverted, nucleotide sequence (see Figure 14-7 of *EMB*). IS elements are simple because they do not encode any genes other than those required for insertion into DNA.

Transposons are more complex. They usually consist of one or more genes flanked by two IS elements. The genes enclosed by the left-flanking and right-flanking ISs may be antibiotic resistance genes, or virtually any other genes. Antibiotic resistance genes have received the most attention because they are medically important and relatively easy to study, since they are carried by bacteria. Figure 14-9 of *EMB* illustrates typical transposons with the flanking ISs prominently shown.

Nevertheless, not all transposons have ISs at either end. Figure 14-10 of *EMB* illustrates an antibiotic-resistance-gene-carrying transposon flanked by inverted repeats. This transposon lacks the double transposase configuration shown in Figure 14-9 Also, in this transposon, the transposase gene (Tnp A) is present only as a single copy.

In summary, two types of transposable elements exist: Simple IS transposons that contain genes required to move from one location to another (e.g., Figure 17-7), and "complex transposons" that contain most of the components found in ISs in addition to several other additional genes. These additional genes can (especially if they confer antibiotic resistance) dramatically change the phenotype of the host organism (e.g., Figures 14-9 and 14-10).

The terms "inverted" and "direct" repeats are confusing! Which are part of the transposon and which are not?

Inverted and direct repeats are shown below, in Figures 1 and 2, respectively. To begin with, a transposon is a segment of double-stranded DNA that contains, by definition, the gene for the enzyme transposase. If it does not contain at least the gene for transposase, then it is not a true transposon. If it is a transposon that transposes by the replicative model (shown in Figure 14-12 of *EMB*), then it will contain the gene for resolvase in addition to the gene for transposase.

Transposons have inverted repeat sequences at their termini. This inverted repeat is transposon DNA. When the transposon inserts into the recipient DNA molecule, it does so at a target sequence. This sequence is duplicated during the process of transposition to form the direct repeat that flanks the transposon. This is illustrated in Figure 14-8 of *EMB*. Note that the direct repeat is in the host DNA, not the transposon DNA.

```
..... A  B  C  D........ D' C' B' A' .....
..... A' B' C' D'........ D  C  B  A  .....
```

Figure 1: Inverted Repeat

```
..... A  B  C  D........ A  B  C  D  .....
..... A' B' C' D'........ A' B' C' D' .....
```

Figure 2: Direct Repeat

How about a walk through Figure 14-12 of *EMB*?

This illustration offers two plausible mechanisms to explain the movement of transposons from one plasmid, or chromosome, to another. Some transposons move via a "conservative" pathway, while others use a "replicative" pathway. Regardless, both of these pathways begin the same way. As Figure 14-12(c) illustrates, a "common intermediate" emerges from cutting and ligating the donor and target molecules.

First, let's track the conservative pathway. Additional cuts are made in the common intermediate that free the transposon from the donor molecule, as shown in Figure 14-12(d). Although the transposon is cut free from the donor, it still remains joined to the recipient. DNA polymerase I then comes into play and fills in the gap in the target molecule (Figure 14-12(e)). The donor molecule is then left open and eventually degraded.

Second, let's focus on the replicative pathway. Here the transposon is duplicated, hence the term "replicative" pathway. The joined common intermediate (Figure 14-12(c)) serves as a template for DNA synthesis. No additional cuts are made, as was the case for the conservative pathway. A cointegrate of the type shown in Figure 14-11 of *EMB* forms. Finally, the enzyme resolvase (e.g., the enzyme coded for by the TnpR gene illustrated in Figure 14-10 of *EMB*) comes on the scene. It drives a recombinational event that results in both the donor and target molecule containing a complete transposon.

Once again: How does the conservative method of transposition differ from the replicative method?

Both methods begin with a donor DNA molecule that contains a transposon, and a recipient DNA molecule that lacks the transposon in question. The main difference between the two pathways lies in the resulting final products. The final product of the conservative method is a recipient molecule that now contains the original transposon. "Original" is the key word here; the transposon in conservative transposition is *not* copied, but is moved intact from the donor molecule to the recipient molecule.

The final products of the replicative method differ. They consist of the donor molecule *and* the recipient molecule, each having its own copy of the transposon in question. It is important to note that the recipient molecule does **not** have a duplicate transposon while the donor still contains the original transposon. This is a common misconception. During replicative transposition, the transposon is replicated and both the donor and recipient DNA molecules receive one of the "daughter" transposons; each has a transposon consisting of an "old" strand and a "new" strand of DNA.

Other minor differences also exist. A structure called a "cointegrate" is an intermediate in the process of replicative transposition, but not in conservative transposition. Also, while both methods require the enzyme transposase, the replicative method also requires the enzyme resolvase.

Key Terms

Plasmid

R plasmid

Col plasmid

DNA modification genes

Conjugation

Self-transmissible

F plasmid

Sex plasmid

Pilus

Hfr

F^+

F'

*ori*T

Insertion sequence

Merodiploid

Tn elements

Transposon

Direct repeat

Inverted repeat

Transposase

Resolvase

Target sequence

Conservative transposition

Replicative transposition

Cointegrate

Methylation

Autonomous

Retrovirus

Provirus

For More Information . . .

Extra help on the topics contained in this chapter is easy to find:

A brief introduction to plasmids and transposons is provided on pp. 273-274 and pp. 284-286 of *Molecular Biology of the Cell*, 3rd edition, by Alberts *et. al.*

Similar introductory comments can be found in *Molecular Biology of the Gene*, 4th edition, by Watson *et. al.*, pp. 202-203. Details of transposons can be found on pp. 332-337 and on p. 336, which illustrates the conservative and replicative models of transposition.

The Wise Owl Says

Uninterpretable results are often the most interesting!

Scientists usually expect a certain outcome from an experiment. That is, experiments are frequently designed to prove or disprove a particular hypothesis. The expected result is usually greeted with only a routine amount of fanfare.

Nevertheless, unexpected results often prove, over the long term, to be more exciting. Unexpected, unforeseen, or uninterpretable results from an experiment stimulate a search for new ideas, novel hypotheses, or brand new models.

Such was the case for plasmids. Initially, scientists were puzzled by the observation that bacteria could become resistant to several antibiotics in a single step. Mutation could be ruled out, chromosomal DNA transfer was considered improbable, and virus transmission was eliminated as a plausible mechanism. Finally, the "uninterpretable" observation of simultaneous acquisition of several gene functions was understood when plasmid DNA was first isolated, characterized, and subsequently demonstrated to confer new properties to host bacteria.

Transposons also represented a strange case of "uninterpretable data." Several changes in maize (corn) plants were noticed as a result of apparent relocations of "controlling genes" to within close proximity of structural genes. For many years, these "movable genes" were poorly understood. In fact, their very existence as classical genes was often questioned, in part because they were originally only recognized in corn plants. "Jumping genes" were eventually discovered in the fruitfly, *Drosophila*. Shortly thereafter, "transposons" were isolated with recombinant DNA technology (see Chapter 15 in *EMB*) and their role in inheritance was firmly established.

Exercise

I. Movement of a bacterial chromosome from an Hfr donor cell to an F^- recipient cell.

Let's make certain we know what Figure 114-4 of *EMB* illustrates. To insure understanding, we will plot some data on a graph. We will then draw a conclusion based on the information in the graph.

Here are the data:

Minutes after culture of Hfr donor cells mixed with F^- represent cells	Percent of F^- recombinants that exhibit donor traits		
	trait A	trait B	trait C
15	10	10	0
30	45	10	0
45	70	40	0
60	80	60	10
75	85	60	20
90	85	65	25

Next plot these data points:

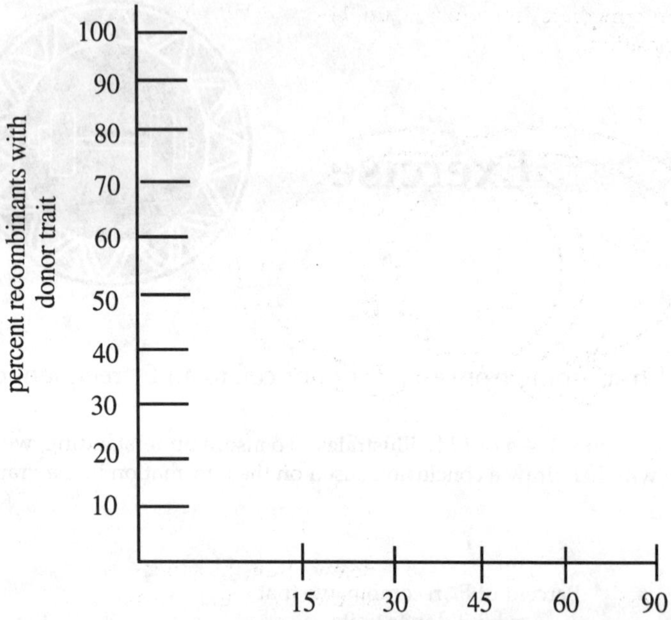

(Please note: The correct graph is illustrated at the end of this chapter.)

Finally, what is your interpretation of the data in the graph on the previous page?

Please note: Using the above conjugation scheme (as illustrated in Figure 14-4 of *EMB*), approximately 1,000 genes have been mapped in *E. coli*!

II. Understanding conservative and replicative transposition.

Beginning with the "common intermediate" shown in Figure 14-6 of *EMB*, let's sketch out the two alternative pathways for transposition.

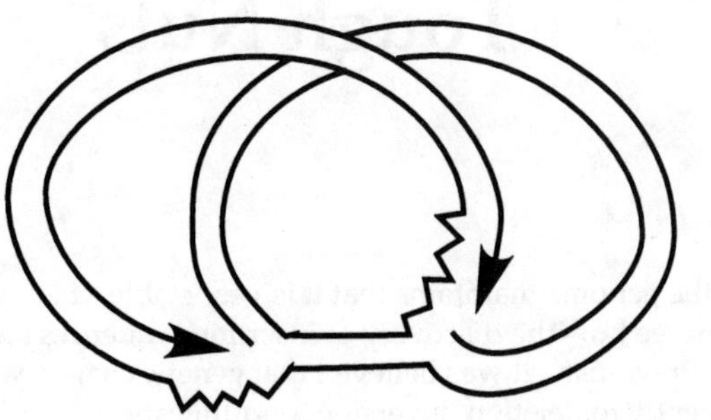

	replicative	***conservative***
a.	**Illustrate below DNA synthesis duplicating the transposon.**	**Below, draw the single-stranded cuts that are made in the intermediate.**
b.	**Draw the final products of recombination below.**	**Show how DNA synthesis is employed to complete the integration of the transposon into the target molecule.**

Please compare part (a.) under replicative with part (b.) under conservative, and vice versa. How different are the molecular events that comprise these steps? How different are the final products?

Tough Nuts

The classical view of the genome maintains that it is very stable, changing very little over time. Consequently, you can see how the discovery of insertion sequences and transposons astonished many molecular biologists! Previously, it was believed that genetic change was due largely to recombination, mutation, and infrequently to deletion, inversion, or duplication.

Insertion sequences and transposons carry the necessary base sequence information for coding the enzymes that accomplish their physical relocation. This makes them remarkably well-equipped for moving around within a genome. Only recently has the virtuosity of movable genetic elements become understood. Even still, several tough questions remain to be answered if we are to understand movable elements in the broader perspective of genetics and their meaning for heredity.

1. How prevalent are movable genes in the biological kingdom? That is, how many types of transposable elements exist in the biological kingdom, or, for that matter, within a single organism? In maize (corn), Ac and Ds transposable elements have thus far been characterized. P-elements have been characterized in *Drosophila* (fruitfly), similar to the Ds elements of corn. In yeast, a Ty element has been discovered.

   Do these well-characterized transposons represent the "tip of the iceberg"? Will, literally, dozens of other transposons, in virtually all members of the biological kingdom, be discovered eventually?

2. Are transposable elements active in "jumping" from one location to another in the human genome? Data generated from the human-genome sequencing project will probably provide evidence for their presence in human chromosomes, should they exist. Nevertheless, once recognized, direct experimentation will undoubtedly be difficult. Initially, cultured mammalian cells (e.g., of human origin) will be the system of choice. Recall, in Chapter 1 of *EMB* , animal cells were described in terms of their usefulness for experimentation.

   It should be kept in mind that gene rearrangement generates antibody diversity. This subject is described in more detail in Chapter 13 of *EMB*. Could the molecular mechanisms involved in those rearrangements represent a similar type of pathway to that employed by transposons as they move from one location to another?

3. Are mobile genetic elements that facilitate evolutionary change becoming more widespread, or are they heading towards extinction? Since the inventory of transposons is incomplete—it is unknown, for example, how prevalent movable genetic elements are in mammals—this question represents a very "tough nut."

Most likely, information for answering each of the above questions will come from the extensive genome sequencing projects now underway. Nematode, fruitfly, plant, and human genomes are being sequenced. As the sequences are analyzed with computer programs designed to recognize the properties expected of movable genes, insights will surely be gained.

Study Questions

1. What would be the result of injecting many copies of the resolvase enzyme into a cell that contains both a Tn3 transposon on its chromosome and an F plasmid with a Tn3 target sequence? Would "neighboring" cells become penicillin-resistant (i.e., acquire an F plasmid with an incorporated Tn3)?

2. What would be the effect of injecting a cell that contains a Tn3 transposon with many copies of the enzyme transposase? Would nearby cells become penicillin-resistant?

3. When transposition occurs by the replicative model, two base sequences are duplicated. What are those two sequences?

4. What is the difference between horizontal and vertical transfer of genetic information? What types of DNA molecules serve as vectors of genetic information in each?

5. Describe in words (without diagrams) how F plasmids transfer from one cell to another. Please be certain to *fully* discuss how certain gene products are involved.

6. Rifampicin is an antibiotic that prevents prokaryotic RNA synthesis. There are several ways in which rifampicin can prevent effective transfer of an F plasmid. What are two?

7. What method of replication is used to replicate a free F plasmid and what method is used in Hfr transformation?

8. What early enzymatic step is needed in Hfr transformation, but not in regular replication?

9. If ColE1 could be altered to contain insertion sequences homologous to sequences in the chromosome and was integrated into the chromosome, would Hfr-like cells arise?

10. When grown on LB medium containing X-gal, Lac^+ bacterial colonies appear purple and Lac^- colonies appear white. When several hundred thousand $F'lac^+/Lac^-$ cells are plated on such medium, the majority of the colonies are purple. However, a few colonies appear that have a well-defined white section to them (the white section is always smaller than the purple section). These are referred to as "sectored" colonies. What is the cause of the this sectoring?

11. If it takes 100 minutes to transfer the entire chromosome of an Hfr cell, when (measuring from time = 0 minutes) are the F genes transferred?

Answers to Chapter 14 Study Questions

1. For neighboring cells to become PCN-resistant, two things must occur: (1) Tn3 must transpose into the F plasmid (catalyzed by transposase and resolvase), and (2) the F plasmid must be transferred into neighboring female cells (catalyzed by the proteins encoded by the *tra* genes). Since resolvase inhibits the expression, at the level of transcription, of the genes for resolvase and transposase, injection into the cell would result in no transposition. Neighboring cells would, therefore, not acquire the transposon.

2. If transposase were injected, the level of transposition would increase and neighboring cells would stand a greater chance of acquiring an F plasmid with the PCN-resistance gene.

3. In replicative transposition both the target sequence of the recipient DNA and the transposon itself are duplicated.

4. Vertical transfer is basically inheritance; the passing on of traits from parent to progeny. Horizontal transfer is the passing of one or more traits from one cell to a neighboring cell, in a sense, from one adult to another. The chromosome is the main DNA molecule passed from parent to progeny (although plasmids might be also), while plasmids are the DNA molecules that can transfer "horizontally" from cell to cell.

5. a. Production of the enzymes necessary for plasmid transfer.
   b. A pilus from F^+ cell brings an F^- cell (the recipient) into contact with the donor (F^+ cell), and a conjugation bridge forms between the two.
   c. One strand of the supercoiled plasmid is nicked at *ori*T by a DNA endonuclease encoded by an F gene.
   d. Modified rolling-circle replication begins (modified, in that initially only the leading strand is replicated). The displaced parental strand is conducted into the recipient cell by certain proteins.
   e. Replication of the parental plasmid DNA is finished in each cell, the donor plasmid DNA is recircularized, and a gyrase enzyme catalyzes the supercoiling of both.

6. The most likely way is by inhibiting transcription of the genes encoding the proteins necessary for transfer. If enough of those proteins were already present, such that plasmid transfer was initiated, rifampicin could block the polymerase from making a primer for the strand that enters the donor cell. Hence, one strand of the plasmid would be transferred normally, although it would remain single-stranded. The parental molecule, however, would be double-stranded, as it does not need a primer.

7. Both free and integrated plasmids (Hfr) are replicated via bidirectional replication. A free or integrated plasmid in the process of transformation to another cell undergoes rolling-circle replication.

8. Nicking of one of the strands of the DNA molecule.

9. ColE1 plasmids could conceivably integrate into the chromosome, but they contain none of the genes needed to transfer themselves from one cell to another.

10. Early in the development of the colony, an F' cell has divided and one of the two daughter cells did not receive a copy of the plasmid. Since that cell would not have had a wild-type copy of the lac gene, it and all its progeny will appear white.

11. Since transfer begins within the integrated F DNA, both the first and last genes to enter the recipient cell are F genes. Therefore, the F genes are transferred at both 0 and 100 minutes.

Concept Map

In order to insure that we comprehend the "big picture," let's prepare a concept map:

Selective Advantage

Plasmids

Transposons

Replication Mechanisms

Include, at least, these major components in your concept map by connecting them with lines and arrows.

- terminal repeat
- conjugation
- insertion sequence
- conservative/replicative transposition
- target sequence
- plasmid replication models
- colicins
- Hfr cell
- direct repeat
- transposase

P.S. Why not add entries from "Key Terms"?

Practical Applications

Genetic engineering (recombinant DNA technology) is leading the way for practical applications of molecular biology. Entire industries have been able to use this technology to produce various pharmaceutical agents and agricultural products. As will be detailed in Chapter 15 of *EMB*, plasmids provide a convenient vector for genetic engineering procedures. The knowledge, developed here in Chapter 14, of plasmid structure, as well as its growth control mechanisms, has been fundamental to designing appropriate plasmid vectors and growth conditions for commercial scale applications.

Transposons have also provided powerful tools for genetic engineers. For example, they can be used as carriers in order to transfer foreign genes into test organisms. The transferred genes can then be examined for their ability to function in the new host organisms and, perhaps, yield new information about gene expression.

Solution for Exercise!:

I.

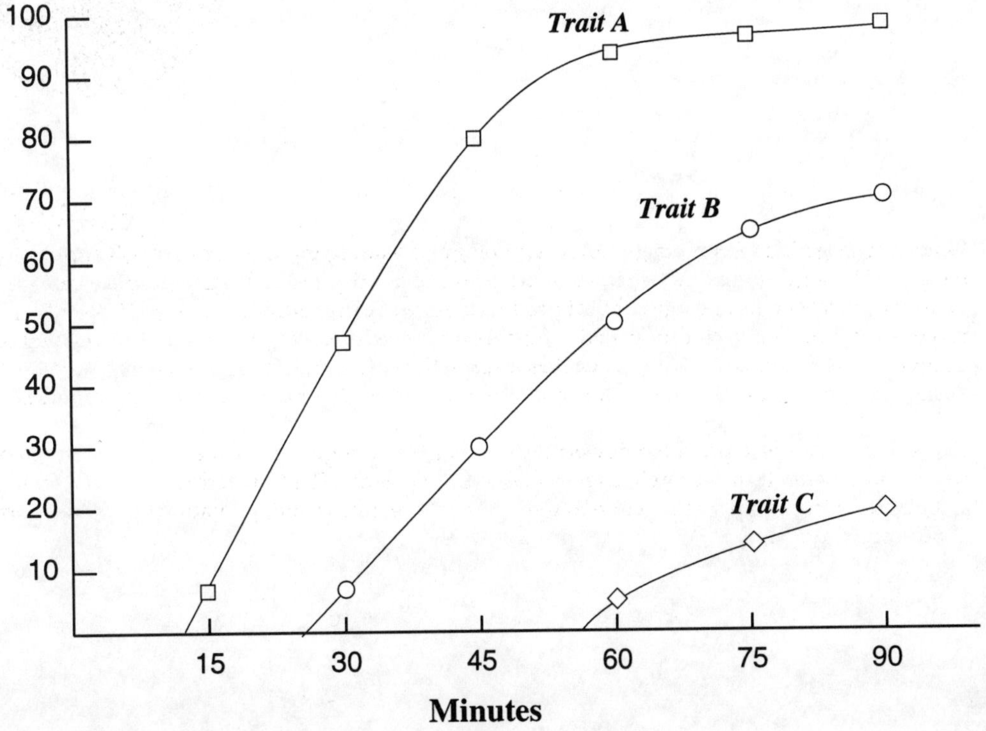

15 Recombinant DNA and Genetic Engineering: Molecular Tailoring of Genes

Here's Help

Why don't we begin with a brief vocabulary lesson?

A "Minimalist" Vocabulary for a Genetic Engineer:

restrict = to digest, or cut, DNA with restriction endonuclease.

vector = a plasmid or bacteriophage used for cloning.

clone = the preparation of multiple, identical copies by replicating a vector in a bacterial host.

genomic library = various large gene fragments, inserted, usually, into a collection of phage or plasmids.

cDNA library = various relatively small cDNAs, inserted, usually, into a collection of plasmids.

screen = a search of a library for one particular gene fragment.

reverse transcriptase = enzyme that synthesizes cDNA from RNA.

probe = a nucleic acid used to identify a target gene fragment via hybridization.

walk = a method of sequential screening of a library, whereby overlapping fragments are used to collect all the pieces that comprise a large gene.

How do restriction enzymes manage to make such highly specific cuts in DNA?

Restriction enzymes are homodimers. This means that they are composed of two identical subunits. These enzymes wrap around the DNA, thus allowing the two enzymatic subunits to interact with both "sides" of the double helix.

The base sequences recognized by these restriction enzymes are usually four to six bases long and often display two-fold symmetry (two regions of identical sequence in a palindromic array). Since each complete turn of the helix contains 10.5 bases, a typical recognition site spans half a turn. The fact that both the recognition site and the enzyme display two-fold symmetry permits a perfect match. Each enzymatic subunit is therefore able to interact with "its own" region of the recognition site. Keep in mind, however, that literally hundreds of different restriction enzymes have been identified, so it is difficult to make rules which apply to <u>all</u> of them!

Here's a review of chromosome walking.

Suppose many copies of a very large DNA molecule are incompletely digested by restriction enzymes so that they are not all cut into identical fragments. That is, cuts are not made at each and every restriction enzyme recognition site on any one particular copy of the molecule. Due to this phenomenon, "overlapping" fragments will be generated such that the sequence at the "beginning" of one fragment is identical to (i.e., "overlaps") the sequence at the "end" of another fragment (see Figure 1 below).

Now please have a look at Figure 2 (below). Suppose the DNA molecule has been restricted (i.e., cut) as shown and the fragments have been cloned. Also, suppose we have isolated a recombinant phage library (or plasmid library) that contains fragment C, and we would like to find other colonies in the library that contain fragments of adjacent portions of the gene (e.g., fragments A, B, D, and E). Since the "beginning" of sequence D is identical to the "end" of sequence C, we could do this by chromosome walking. A radioactive cDNA probe, complementary to the sequence at the end of C, could be used to screen the library. That probe would hybridize with sequence D. (*Remember, in screening a library, the pattern of phage plaques (or plasmid-containing bacterial colonies) is transferred to a nitrocellulose filter, the phage proteins are dissolved, and the filter is incubated (i.e., hybridized) with a solution of the nucleic acid probe. A variation of this approach, first mentioned in Chapter 3 (Figure 3-12), is illustrated in Figure 15-8 of* EMB.) To "walk" further down the molecule, next we would make a probe complementary with the end of fragment D and screen the library to isolate fragment E. We could also walk in the other direction by beginning with a probe complementary to the other end (i.e., the "beginning") of C. This would be helpful in isolating clones of fragments B and A.

ABCDEFGHIJKLM

JKLMNOPQRSTUVWXYZ

Figure. 1. Here is an example of overlapping fragments. Note that the sequence at the end of the upper restriction fragment is identical to that at the beginning of the lower. Therefore, a probe complementary to the right-hand end of the upper fragment would also be complementary to the beginning of the lower fragment.

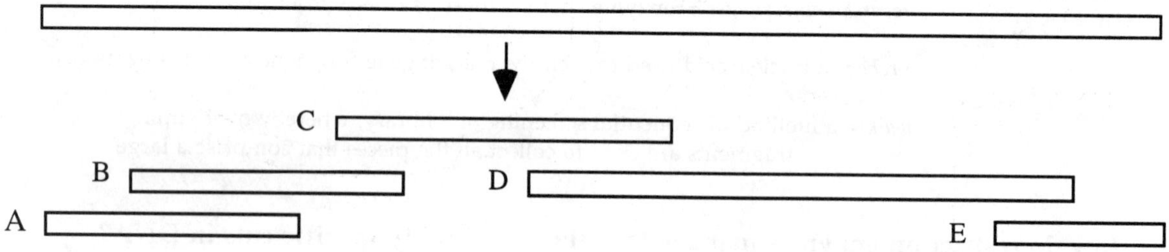

Figure. 2. Many copies of the top DNA molecule have been incompletely digested with restriction enzymes to yield overlapping fragments A-E.

Here is a brief review of PCR:

Let's go step by step through the polymerase chain reaction illustrated in Figure 15-5 of *EMB*. Let's start with a single double-stranded DNA molecule that contains the target sequence you want to amplify, and let's pretend we are running a PCR reaction!

a. Heat a mixture of primers, nucleotides, and DNA so that the double-stranded DNA molecule denatures.

b. Now, cool the mixture. This permits the primers to hybridize to their complementary sequences at the 3' end of each single-stranded target sequence.

c. After providing sufficient time for primers to attach, add DNA polymerase, and raise the temperature slightly so the enzyme can function efficiently. DNA polymerase will begin synthesizing DNA at the 3' end of each primer, making the rest of the molecule double-stranded.

d. After DNA synthesis has occurred, heat the mixture in order to denature the double-stranded molecules. Remember, each round of PCR doubles the amount of DNA present at the beginning of the round. The cycle is repeated beginning with step a., followed by b., etc.

How does "bidirectional" cloning, mentioned in Table 15-2 of *EMB* work?

Using a single cutting enzyme, such as EcoRI, to cut both the plasmid and the piece of DNA to be cloned, cohesive, complementary (sticky) ends are generated. The insert (i.e., foreign DNA) and the plasmid anneal, and, with the addition of a DNA ligase enzyme, they are sealed together.

This is perhaps the simplest cloning procedure, because only one cutting enzyme is employed both to cut the DNA fragment to be inserted and to open the plasmid vector. This simple (single cutting) enzyme results, however, in two possible orientations of the DNA fragment in the plasmid vector. For many applications in genetic engineering, the "bidirectional" orientation is suitable.

What is meant by "bidirectional" cloning?

Simply stated, a DNA fragment can insert into a (plasmid) vector either left to right, or right to left. That is, the fragment can insert in either of two directions. That's why it's called "bidirectional."

Perhaps the best way to understand it is to diagram it out. Shall we do so?

Recall, here is the question we will answer by diagramming it out: What does the term "bidirectional insertion" of a DNA fragment mean?

Let's begin with the two components, as illustrated below:

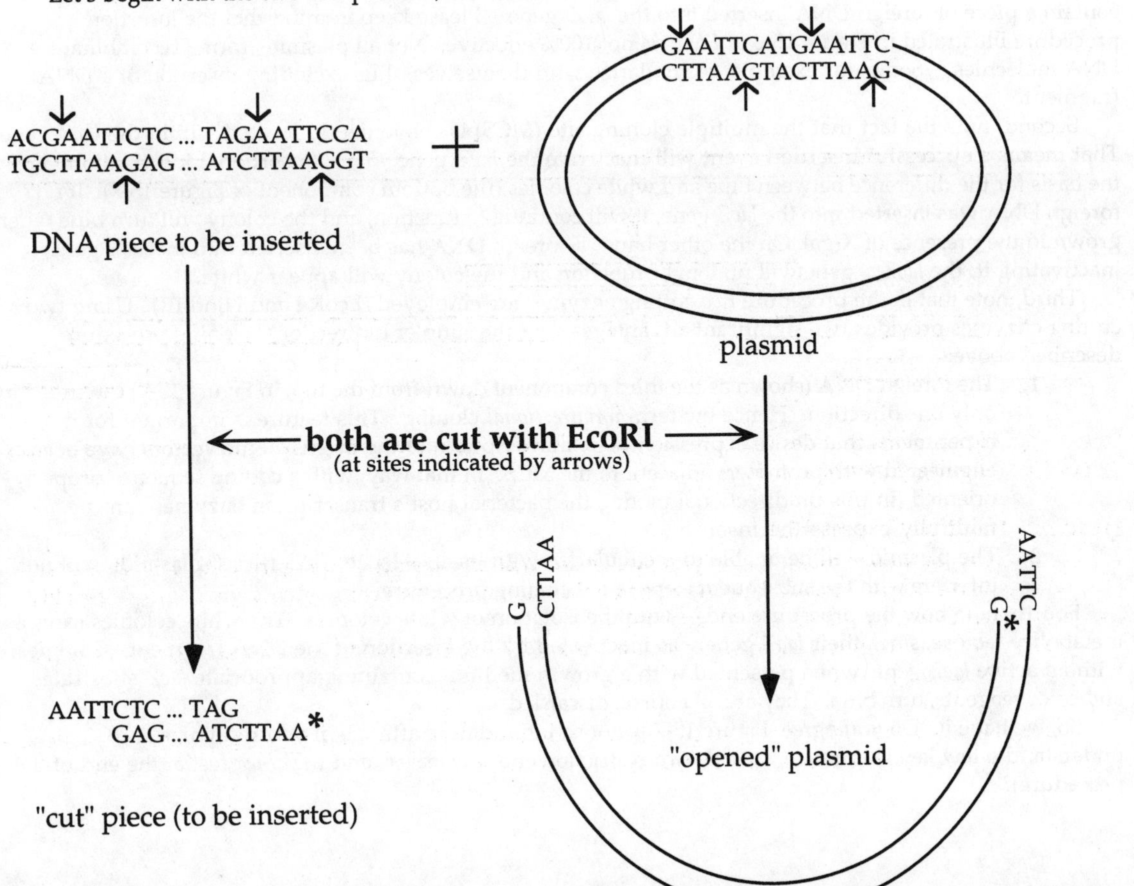

Here's the answer: The cut piece can "fit" into the opened plasmid in two orientations. One orientation has the two asterisks adjacent to each other.

The other orientation has the cut piece "flip-flopped" and "upside-down," so that the asterisks are as far away from one another as possible.

Because the cut piece can fit into the opened plasmid in those two very different orientations, this type of insertion, in which a single restriction endonuclease is used to cut both the DNA sample and the plasmid, is called "bidirectional." The simpler cloning methods based upon this procedure are referred to as bidirectional cloning. That is, although the method is called "cloning," it really refers to the insertion process, rather than the cloning process.

It is suitable for most applications for which the desired end product is multiple copies of the starting material.

However, for more sophisticated applications, directional cloning (Figure 15-7 in *EMB*) is employed. We will now describe that method.

How about a review of "directional" cloning (Figure 15-7 of *EMB*)?

Let's review that illustration step by step, pointing out the key elements as we go.

First, note the plasmid's *Ap*^{r} and *lacZ* genes. The *Ap*^{r} gene will be important for selecting from the millions of bacteria in the "infection mixture" those bacteria which were successfully infected with a plasmid. The way it works is simple: If a bacterium is infected with a plasmid, the bacterium becomes resistant to the antibiotic ampicillin. That is, uninfected bacteria are *unable to grow in the presence of ampicillin,* and are eliminated from the population. It is important to include this step in a cloning procedure, because not all plasmid infections are 100% successful. Lots of bacteria end up without a plasmid!

What's the value of the *lacZ* gene? It will be useful for identifying from population of *all* plasmid-containing bacteria ("selected for" with the ampicillin treatment just mentioned) those bacteria which contain a piece of foreign DNA inserted into the *lacZ* gene. Please keep in mind that the insertion procedure illustrated in Figure 15-6 of *EMB* is not 100% effective. Not all plasmids form "recombinant DNA molecules." Some, for example, recircularize with themselves, thus excluding insertion of a DNA fragment.

Second, note the fact that the multiple cloning site (MCS) is conveniently located within the *lacZ* gene. That means a successful insertion event will inactivate the *lacZ* gene. This "insertional inactivation" is the basis for the difference between blue and white colonies (the bottom component of Figure 15-7). If no foreign DNA was inserted into the *lacZ* gene, it will continue to function, and the colony will turn blue when grown in the presence of X-gal. On the other hand, if foreign DNA has been inserted into the *lacZ* gene, inactivating it, the *lacZ* gene will no longer function and the colony will appear white.

Third, note that in the procedure *two* cutting enzymes are employed (EcoR I and Hind III). Using two cutting enzymes provides two significant advantages over the simpler bidirectional cloning procedure described above:

1. The foreign DNA (shown as the third component down from the top, in Figure 15-7) can insert in only one direction. Hence the term *unidirectional* cloning. This feature is important for experiments that desire expression (e.g., transcription) of the insert. Some vectors have been engineered with promoters adjacent to the MCS. In that way, with a coding sequence properly oriented (in this unidirectional mode), the bacterial host's transcription enzymes can faithfully express the insert.
2. The plasmid will be unable to recircularize with itself. Hence, "insertless" plasmids will not interfere with the subsequent steps in the cloning procedure.

Fourth, note how the procedure ends—with the isolation of white colonies. The white colonies cannot metabolize lactose, since their *lacZ* gene was inactivated by the insertion of the DNA fragment. Colonies with an active *lacZ* gene, when presented with a growth medium containing appropriate *lacZ* substrates and color reagents, turn blue. They are, of course, discarded.

So, we have it. Do you agree, Figure 15-7 is not so intimidating after all, if you recognize and understand a few key features (e.g., *lacZ*, two restriction endonucleases, and the color test at the end of the procedure)?

Why is the detection of recombinant molecules such an important step in the cloning procedure?

EMB lists three possible outcomes of an attempt to insert a DNA fragment into a vector. Brief review of those outcomes explains why only a fraction of the treated plasmids will have useful inserts.

In addition, during the transformation of bacteria with plasmids (please see Figure 15-6 (third illustration down), of *EMB*) the efficiency is not 100%. Many bacteria will not be transformed. It is, therefore, necessary to eliminate those untransformed bacteria from the experimental procedure. This is where the ap^r trait comes into play (please see Figure 15-7 (top illustration) of *EMB*). That trait, carried on the plasmid, permits the scientist to screen out (i.e., eliminate) bacteria that have not been successfully transformed. How? By treating the "transformed" population of bacteria with the antibiotic ampicillin. It will kill all the non-transformed bacteria, since bacteria, normally being sensitive to this antibiotic, gain resistance to it only by virtue of having been transformed by the ap^r plasmid.

Why do so many cloning procedures begin with mRNA?

If one wants to clone the coding sequence of a particular gene, one could isolate and clone the gene itself, or one could isolate mRNA transcribed from that gene, make cDNA copies of it, and clone the cDNA. The enzyme "reverse transcriptase," would be employed. There are several reasons why cloning procedures often begin with mRNA. One reason is availability of multiple copies of mRNA. Even a cell that produces lots of gene product will probably have only two copies of the gene itself. Nonetheless, it will have *many copies* of the mRNA from that gene. This is just one reason to begin with mRNA. Remember, if one chooses to begin with DNA, one must prepare restriction fragments of the genome and clone them into bacteria to make a genomic library. Then one must find the colony in that library that contains the fragment with the gene of interest (screen). The biggest problem with this method is that the fragment which is finally isolated may contain only a fraction of the gene, since restriction enzymes may have made a cut within the gene. Thus, beginning with RNA might be more desirable.

In order to begin with mRNA, one needs to identify cells that have a high level of expression of the gene and isolate mRNA from those cells. This method follows the assumption that most, or at least a large portion, of the highly expressed mRNA will belong to the gene of interest. Using reverse transcriptase to make cDNA from the mRNA and cloning the cDNA would insure that the coding sequence is cloned in its entirety. *EMB* gives examples (ovalbumin from the chicken (oviduct) and CO_2-fixing enzymes from plants).

Yet another factor must be considered if one is thinking about beginning with DNA: Is the gene of interest from a eukaryotic organism? If the overall goal of cloning the gene is to observe its expression in a bacterial system, then cloning the gene itself presents problems. Recall, a bacterial genome contains no introns, while that of a eukaryote does. These introns must be removed from the primary RNA transcript for a functional product to be made. Since the genes of bacteria do not have these introns, bacteria are not equipped to deal with their presence in a cloned eukaryotic gene. There are also differences between prokaryotic and eukaryotic promoters, so the bacterial transcription "machinery" may not even be able to transcribe a particular eukaryotic gene. By cloning cDNA (made from a mature mRNA molecule with a contiguous coding sequence), the problem with introns is circumvented, and integration near a prokaryotic or viral promoter allows one to achieve expression within a bacterial cell.

Key Terms

Genetic engineering

Plasmid DNA

Restriction enzymes

Vector

Palindrome

Cohesive ends

Restriction map

Genomic library

PCR

Shuttle vectors

Directional cloning

Cloning

Selection

Screening

Selectable marker

Reverse transcription

cDNA

Insertional inactivation

Alpha-complementation

For More Information . . .

Additional explanations on these topics are readily available:

Excellent coverage of most of the topics in this chapter can be found on pp. pp. 291-305, 308-317, 323-324, and 327-328 in Alberts *et. al., Molecular Biology of the Cell,* 3rd edition.

Restriction enzymes and genomic libraries are described in pp. 266-274 in Watson *et. al., Molecular Biology of the Gene,* Vol. I, 4th edition. Menlo Park, California: Benjamin-Cummings Publishing Co., 1987.

The Wise Owl Says

We are now in the golden age of biology!

The relative importance of various academic disciplines rises and falls over the course of years, decades, or even centuries. The discipline of molecular biology started gaining momentum with the elucidation of the structure of DNA and with follow-up studies that explained how genetic information is converted to protein structure.

A major boost in the productivity of molecular biology research began when restriction endonucleases, the enzymes that cut DNA at specific sites. were discovered. These enzymes provided essential tools for gene cloning and mapping.

Gene manipulation technology has been the driving force of this golden era. Virtually all areas of biology, including genetics, physiology, agriculture, taxonomy, biochemistry, evolutionary biology, gene therapy, and even forensic medicine are being propelled to new levels of understanding with recombinant DNA technology.

Exercise

I. Beginning with a sample of hemoglobin mRNA, let's design a procedure for cloning the human hemoglobin gene. We will go step by step, drawing in the essential components. In part A, we will prepare a probe, part B the library, and part C the identification of clones.

A. Preparation of the detection probe for screening this genomic library, to be prepared in part C.

______________________________ hemoglobin mRNA (starting material)

\+

name of enzyme: ____________________

↓

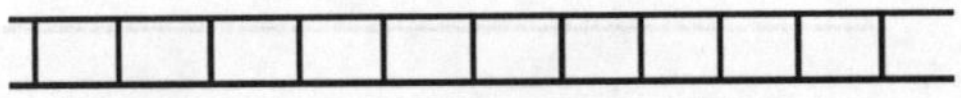

dsDNA copy (i.e., cDNA)

\+

plasmid

↓

explain how the cDNA is inserted into the plasmid:

____

____

____

____

diagram the plasmid with its insert:

explain how the cDNA plasmid is selected: ____________________

____

____

____

multiple copies of "hemoglobin" plasmid

explain how the insert is isolated and radioactively labeled:

____

____

____

____

radioactive hemoglobin cDNA

B. Preparation of the genomic library.

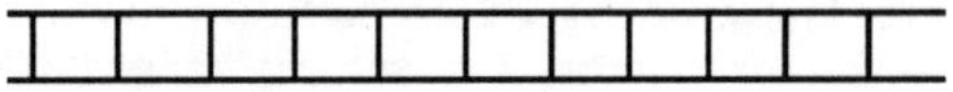

human dsDNA (starting material)

explain how the DNA is cut into pieces:

____

____

____

____

fragments:

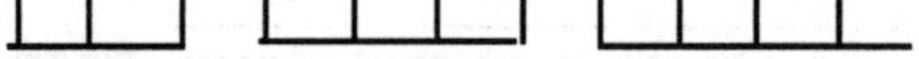

+

phage

explain how the fragments are inserted into the phage:

____

____

____

____

diagram the phage
with packaged
recombinant DNA:

explain how the phage is grown:

____

____

____

____

multiple copies of the phage

C. Identification of clones that contain fragments of the hemoglobin gene.

plating of phage ⟶ plaques

explain how the radioactive cDNA from
part A would be employed:

____

____

____

____

identification of plaques that
represent clones containing
hemoglobin gene fragments

II. Having isolated a set of clones that each contain a fragment of the hemoglobin gene, how would one go about lining up the fragments (5'P → 3'OH direction) in order to display the overlapping pieces? Answer: Prepare restriction maps of each clone and line them up using the regions of overlap for orientation.

How about some practice lining up restriction map fragments? Here are two sample exercises:

1. A 1500 bp gene generates the following restriction fragments:

with Sal I:	400* bp and 1100 bp
with Hind III:	600 bp and 900 bp
double digest:	400, 500, 600 bp

*Tests demonstrate that this fragment is at the 3' end.

Draw the map (answer at the end of this chapter):

2. A 1400 bp gene generates the following restriction fragments:

with Pst I:	330 and 1070* bp
with Bgl II:	600 and 800 bp
double digest:	330, 470, 600 bp

*Tests indicate that this fragment is at the 3' end.

Draw the map (answer at the end of this chapter):

III. Here's a really high-tech, but practical, exercise! Let's alter the nucleotide sequence of a cloned gene and determine whether the protein product exhibits improved temperature tolerance. Assume the cloned gene is an enzyme used as an ingredient in laundry detergents. By improving its temperature tolerance, it will work better in the hot-water cycle of an automatic clothes washer.

First: Obtain the complete nucleotide sequence of the genetic code.

How would you sequence the gene?
To do this, would you begin with a genomic clone, or a cDNA clone?

Second: Select a region of the sequence to alter.

Would you choose an internal or a terminal region?

Third: Remove the chosen region and replace it with a synthetic fragment that encodes a different amino acid sequence.

How would you remove the sequence?
How would you synthesize a replacement?

Fourth: Express the new gene and test the protein product for improved temperature resistance.

How would you induce expression of the new gene?

You now know how to custom design proteins! Neat, eh?

Far Out

The PCR procedure (see Figure 15-5 of *EMB*) has been sufficiently developed to be used routinely in some clinical laboratories for diagnostic procedures. It is employed as a rapid, but expensive, method for detecting the HIV (AIDS causing) virus in patients who suspect they may be carrying the virus, but have not yet shown any overt symptoms of infection. Primers complementary to some of the virus's nucleotide sequences can be employed for use in a PCR reaction of this type.

As is always the case with the PCR procedure, only tiny amounts of DNA are required for a successful test. This virtue of the PCR procedure can be exploited for prenatal diagnoses of human embryos. A small amount of embryonic tissue can be collected and its DNA extracted. The DNA can then be tested for disease-causing genes by using appropriate PCR primers. Therefore, a fetus that does carry a genetic disease can be identified early in pregnancy.

Here's where we begin to go "far out." Because a tiny amount of DNA, such as the amount found in a single human embryo cell, is sufficient to drive a PCR procedure, why not pre-test embryos for disease-causing genes before implanting them in the mother's uterus? By injecting the prospective mother with appropriate hormones, she will be induced to ovulate multiple eggs. The eggs can then be collected, fertilized, and grown to the 8-cell stage. At this time, a single cell can be microsurgically removed from the embryo without impairing its ability to develop into a normal and healthy fetus. The DNA can be extracted from the single cell and tested using the PCR procedure.

Suppose 10 eggs are fertilized at once, and all 10 of the 8-celled embryos are individually tested for various disease-causing genes with appropriate primers in a PCR procedure. A decision could then be made concerning which of the 10 eggs is "best" and should be implanted into the mother's uterus. Once implanted, the child would develop through a normal pregnancy and be born with its genetic profile already known. The "far out" aspect again comes into play when other genes, such as those that determine hair color, eye color, and other features, are also analyzed in the 8-celled embryo. A prospective mother could conceivably sort through several, even hundreds, of embryos before selecting one for implantation. Combining the testing of eggs with the use of sperm banks would provide an effective system to select for or against particular sets of features in human offspring.

Is this too "far out"? Should such technologies be developed? Or, would society be better off without it?

Study Questions

1. Restriction endonucleases make one of two types of cuts in the DNA fragment on which they act. What are the natures of these two types of cuts?

2. The recognition site of the restriction enzyme Hpa I is GTTAAC (one strand shown). Assuming that any base has an equal chance of occurring at any given point in the genome, how many Hpa I recognition sites would be present in a DNA molecule containing 40960 base pairs?

3. Suppose you are given a virus that has 7 recognition sites for the restriction endonuclease HindIII in its DNA. Also, suppose that you have cDNA made from the mRNA of a gene that codes for a protein in the virus's coat. How would you use Southern blotting (described on pp. 178-179 of *EMB*) to find the HindIII fragment that contains this gene?

4. If a restriction enzyme cuts at a sequence of only 4 to 6 bases, what prevents the restriction enzymes within a bacterium from cutting the bacterial DNA?

5. Let's assume that you have isolated a prokaryotic gene that codes for a useful enzyme. How would you clone this gene into *E. coli* in a manner that would allow the gene to be replicated?

6. What is the advantage of using a cloning vector that contains two different genes for antibiotic resistance?

7. ColE1 plasmids (described in Chapter 14 of *EMB*) contain genes that code for colicins. These proteins are secreted by the producing cell and kill non-colicin producing cells. How might this fact be exploited in recombinant DNA research?

8. Suppose that a plasmid that carries two genes for antibiotic resistance, tet^r and amp^r, is restricted with the enzyme EcoRI. The only EcoRI recognition site on the plasmid is within the tet^r gene. DNA from the frog *Xenopus* is cut with the same enzyme, and the fragments are allowed to anneal with the digested plasmids. Bacteria are then transformed. What are all the possible antibiotic resistance phenotypes of the bacteria in the culture? How would you select for bacteria that have acquired recombinant plasmids?

9. The size of the product formed when a restriction digest is annealed and ligated varies according to the DNA concentration in the annealing mixture. Why?

10. In eukaryotes, what is the difference between a gene and cDNA made from the mRNA transcribed from that gene? What about in prokaryotes?

11. The following segment of DNA has been cut in four places by a restriction enzyme:

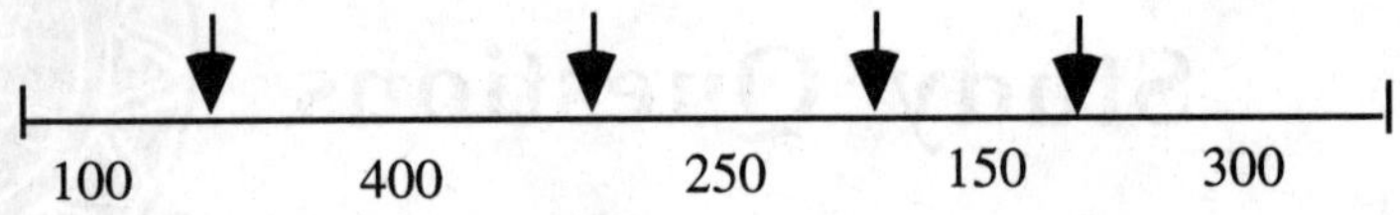

Numbers indicate the size of fragments (in kilobases).

How many bands would the digestion products produce after gel electrophoresis, assuming that each and every segment of DNA was cut as shown above (i.e., total digest)? Suppose that in the sample of DNA each and every segment was not fully cut (i.e., partial digest), how many bands (i.e., digestion products) would be observed after gel electrophoresis?

Answers to Chapter 15 Study Questions

1. Some restriction enzymes produce flush (blunt) ends, in which single-strand cuts are directly opposite one another. Other restriction enzymes generate cohesive ends, in which single-strand cuts are staggered by several nucleotides (Please see Tables 13-1 and 13-2 in *EMB*).

2. 10: $40960 \times (1/4)^6$

3. First, restrict the viral DNA with HindIII and separate the fragments via gel electrophoresis. Then, under denaturing conditions, transfer the bands to a nitrocellulose filter. Use a labeled probe, made from the cDNA, to visualize the band to which it hybridizes. This band contains the gene of interest.

4. In a bacterium's DNA, sites recognized by the restriction enzymes of that bacterium are usually methylated on 1 base. This is a defense mechanism to prevent self-destruction.

5. First, treat the DNA with a restriction enzyme that cuts on either side of the gene. Then, restrict a plasmid containing a selectable marker (e.g., a gene conferring antibiotic resistance) with the same enzyme and mix the plasmid with the gene to be cloned. This should be done under conditions that will allow the sticky ends of the gene to anneal with those of the plasmid. Finally, introduce these recombinant molecules to a culture of *E. coli* and select for successful transformants.

6. If two genes for antibiotic resistance are available, then the foreign DNA can be cloned into one of them. The advantage of this is that it allows selection, not only for bacteria that have acquired the vector, but also for bacteria that have acquired vectors containing the foreign DNA. (For an example of how this is done, please review study question 8.)

7. Many potential host bacteria fail to take up a vector during the transformation process. Those bacteria that lack a ColE1 plasmid will be quickly eliminated from the population by the action of the encoded colicins produced by the bacteria that did successfully obtain a functional plasmid during the transformation procedure.

8. There would be three possible antibiotic-resistance phenotypes. Bacteria that had not acquired a plasmid would be sensitive to both *amp* and *tet*. Bacteria that had acquired a non-recombinant plasmid would be resistant to both *amp* and *tet*. Bacteria that had been transformed with a plasmid containing a *Xenopus* insert would be resistant to *amp* but sensitive to *tet*. To select for bacteria with a recombinant plasmid, first plate the bacteria on medium containing ampicillin, then replica-plate those colonies to medium with both ampicillin and tetracycline.

9. At low DNA concentrations, simple, circular fragments form. At high DNA concentrations, multiple fragments anneal to form dimers, trimers, etc.

10. The eukaryotic gene contains both the coding sequence (exons) and introns (see Chapter 9 in *EMB*). The cDNA, however, contains only the coding sequence. Since prokaryotic genes contain no introns, the gene is the same as the cDNA.

11. Total digest = 5; partial digest = between 2 and 13, depending on the extent of digestion. Keep in mind that digestion products of the same size, regardless of whether they are cut from different locations in the DNA sample, will co-migrate during gel electrophoresis, giving a single band. This is the case, for example, with the 400 bp fragments produced by partial digestion.

Concept Map

A concept map might be just the ticket for understanding the tools employed by genetic engineers:

Restriction Enzyme

Insertional Inactivation

Vector

Restriction Map

Colony Hybridization

Genomic Library

Chromosome Walking

PCR

Please include the following terms in your concept map:

plasmid	palindrome
cohesive ends	reverse transcriptase
cDNA	phage

<u>P.S.: Please add additional terms from Chapter 15!</u>

Solution for Exercise!:

II.

1.

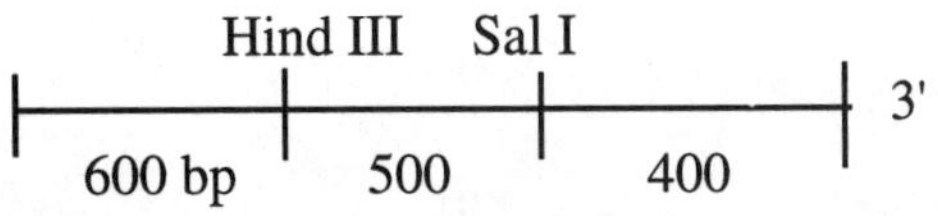

2.

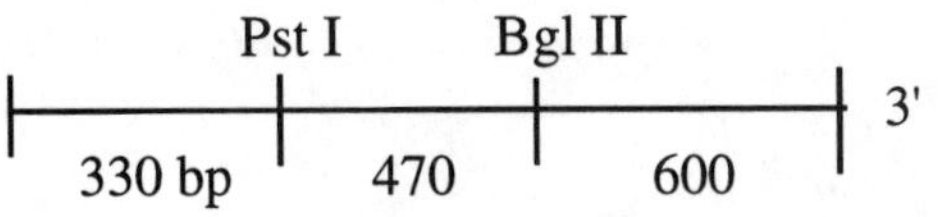

III. First: Begin with a cDNA clone, since it lacks introns. See Chapter 3 in both *EMB* and this manual for sequencing methods.

Second: This is a tough call. Perhaps a region that codes for charged amino acids or hydrophobic amino acids would have the most significant effect on protein folding. See Chapter 4 in *EMB*, especially Figure 4-6.

Third: Remove the fragment by using restriction endonucleases (consult the restriction maps you prepared when the gene was originally cloned). Synthesize a replacement fragment with the method explained in Chapter 3 of *EMB* (Figure 3-21).

Fourth: Construct a new gene containing the coding sequence for the enzyme and a bacterial promoter. Insert this construction into a plasmid to be expressed in a suitable bacterial host. Finally, harvest the enzyme and test its properties.

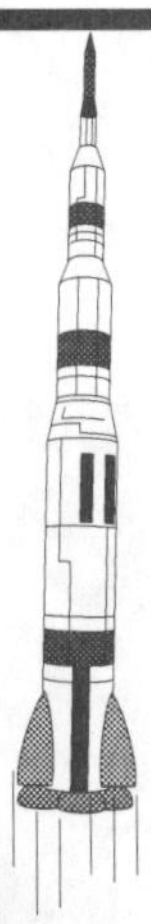

16

Molecular Biology in Expanding its Reach

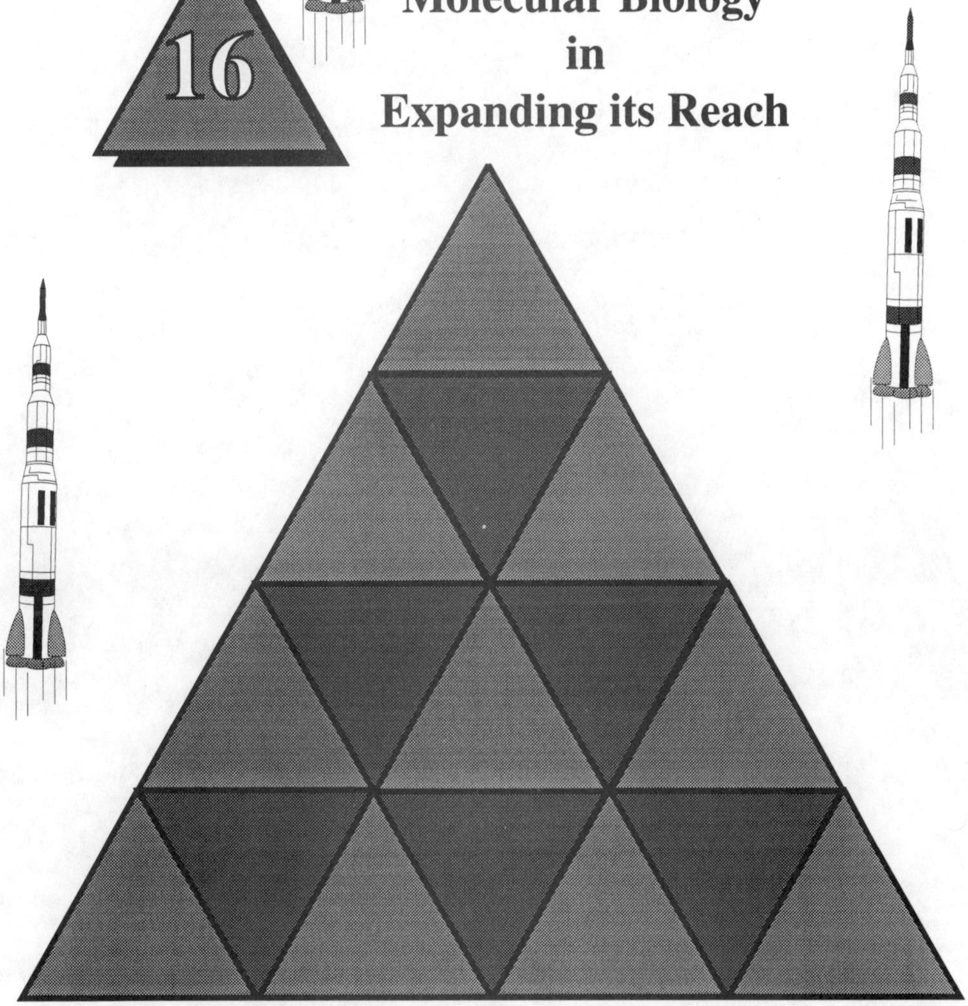

Here's Help

What do we mean by "site-specific mutagenesis"?

In simplest terms, "site-specific mutagenesis" (or *"site-directed mutagenesis"*) refers to a technique that scientists can use to make specific, selected alterations in the base sequence of a gene. Unlike traditional mutagenesis, this technique allows you to alter predetermined regions of the gene and study the effects of such alterations on the activity of the gene or its gene product.

The gene to be modified is first cloned into a cloning vector such as the M13 phage, which has a single-stranded (+) DNA genome. Why do we use the M13 phage? It is tailor-made for this type of site-specific mutagenesis experiment! M13's single-stranded genome is normally converted to a double-stranded replicative form (RF) during phage DNA replication. Then the (-) strand of this RF molecule is used as a template for the synthesis of additional copies of the (+) strand. Thus we can clone our gene into the double-stranded version of the M13 chromosome, use this recombinant phage DNA to transform *E. coli* cells, and isolate the single-stranded copies of our recombinant molecule from the M13 phage progeny.

How do we carry out the actual site-specific mutagenesis? Using an automated DNA synthesizer, we synthesize a short oligonucleotide that is nearly-complementary to the region of the gene that we want to modify, *except that this oligonucleotide contains the altered base sequence that we wish to introduce into our gene.* We allow this synthetic oligonucleotide to anneal with the single-stranded M13 chromosome that contains our gene of interest. There it can serve as a primer for the *in vitro* synthesis of double-stranded M13 phage DNA. The new double-stranded phage DNA is once again used to transform *E. coli* cells, but this time the progeny phage will contain the *mutant* gene in their single-stranded DNA! Once we have copies of the altered gene available, we can re-clone them into an appropriate expression vector and introduce the altered gene into a host in which we can study effects of our site-specific mutagenesis.

Here's a brief summary of the process involved in producing transgenic mice.

1. First, we obtain embryonic stem (ES) cells from an early-stage mouse embryo called a blastocyst. At this stage of embryonic development, the cells have not yet differentiated. We generally use ES cells that are homozygous for an easily-observable phenotypic trait such as dark coat color in order to simplify the process of identifying offspring which ultimately develop from these cells.
2. The ES cells are grown in culture, each original ES cell giving rise to a clone of identical cells.
3. The cloned cells are transformed, using a genetically-engineered gene of interest. This gene is linked to a second gene, such as a promoterless neomycin resistance *(neo^r)* gene, which we can use to select for ES cells that have undergone transformation.
4. The transformed cells are allowed to grow for a period of time prior to selection, to allow the new gene to be incorporated into the genome of the ES cell via homologous recombination.
5. Next, in order to select for cells that have integrated the new gene into their genomes, the old culture medium is replaced with fresh medium containing a neomycin-like substance called G418. Only neomycin-resistant cells will be able to grow in this medium, namely those that have integrated the neo^r gene present on the recombinant DNA into their genome. Since the neo^r gene lacked a promoter of its own, *it will only be expressed if it is situated next to a promoter that permits its expression,* in this case the promoter of our foreign gene.
6. To verify that the neo^r cells did in fact contain our gene of interest, we would perform Southern blotting, using DNA of the new gene as a probe. Neomycin-resistant cells which had undergone successful transformation with our foreign DNA should exhibit *two* bands: one corresponding to the copy of the gene

originally present in the ES cell DNA, the second, heavier band corresponding to the new copy of the gene linked to the *neo*^{r} gene.

7. Now, time to produce some transgenic mice! The transformed ES cells are injected into a blastocyst from a different mouse. In order to simplify identification of transgenic baby mice, this second blastocyst is homozygous for a dissimilar trait. For example, if the transformed blastocyst cells were homozygous for the gene responsible for black fur, the blastocyst into which the cells were injected might be homozygous for white fur. If all goes as planned, these transplanted ES cells will become incorporated into the developing embryo, where they will become part of all of its tissues and organs.

8. The blastocysts are implanted into the uterus of a pseudopregnant female mouse (e.g., she is not actually pregnant, but her body is hormonally ready to accept the embryos), where they complete their development. With a little bit of luck, the female mouse will give birth to a litter of baby mice, some of which will be transgenic. We can identify the transgenic babies by looking for those individuals that are chimeric ("mixed")for the phenotypic trait (such as coat color) that differed in donor ES cells and recipient blastocyst. In this case, the mice might have patches of black fur interspersed with the white fur.

9. After allowing these transgenic mice to mature, the males are mated with female *tester mice* (in this case, female mice with white fur), and the baby mice are examined for the presence of the trait (black fur) contributed by the transgenic ES cells. The rationale is as follows: since transgenic ES cells can be incorporated into *any and all* tissues of the developing blastocyst, some of these transgenic cells should end up in the reproductive organs of the baby mouse. When that chimeric baby mouse matures and reproduces, some of its gametes (sperm or eggs) will contain both the gene responsible for white fur and the new transgene. If a sperm containing this gene combination fertilizes an eggs from a homozygous, non-transgenic white mouse, among the offspring we should find some transgenic baby mice with solid black fur.

10. Test the DNA of the transgenic mice by Southern blotting to verify that the new gene of interest is present. Since these mice are *heterozygous* for both the transgene and coat color, selective breeding between transgenic males and females will produce approximately 25% homozygous transgenic offspring.

What are restriction fragment length polymorphisms (RFLPs) ?

Recall that a mutation is a heritable change in the genetic material of an organism. By altering the base sequence of a gene, a restriction site located within or near that gene may also be lost or gained. These differing patterns of restriction fragments that result from the loss or gain of restriction sites in the DNA molecule are referred to as **restriction length polymorphisms** or **RFLPs.**

Why are RFLPs so useful in the diagnosis of genetic disorders?

If we discover that a particular RFLP pattern is associated with the allele responsible for a certain genetic disorder, then we can use that RFLP pattern to identify individuals who possess that defective allele. Just as a person may be homozygous or heterozygous for a particular gene, so may they be homozygous or heterozygous for the DNA sequences which make up that gene. RFLP patterns in heterozygous individuals are a combination of restriction patterns produced by each of the individual alleles present. Thus, if allele A produces two DNA fragments, 1.0 and 2.0 kb in size, when cut with *EcoRI* , and allele B of the same gene produces only a 3.0 kb DNA fragment, then an individual with the genotype AB would exhibit a combination of the two restriction patterns, and their DNA would exhibit bands at 1.0, 2.0 and 3.0 kb. This information can be extremely valuable under a variety of different circumstances. A person can learn whether they are an asymptomatic carrier of a recessive genetic disorder, and therefore at risk of transmitting the gene to a future child. In the event that both parents are heterozygous, they can learn whether their unborn child will inherit the genetic disorder. In addition, certain dominant genetic disorders such as Alzheimer's disease and Huntington's disease do not begin to exhibit their symptoms until the latter part of a person's life. RFLP testing can enable a person to learn whether they have inherited the gene responsible for these disorders long before the first symptoms appear.

What is the difference between RFLPs and VNTRs?

VNTRs are actually a special class of RFLPs. Unlike the RFLPs described in the preceding section, which are associated with specific alleles of a gene, VNTRs or **variable number tandem repeats** arise from the presence of *different numbers of copies* of short, tandemly-repeated base sequences at various locations throughout the genome. VNTRs exhibit much more diversity that the RFLPs associated with traditional genes. Most individuals possess only two alleles of a gene, and the number of different alleles of a gene present within populations is relatively small. In contrast, VNTRs may exist in many different locations within the genome, and at each location, the number of copies of the repetitive sequence present may range from one to many. Thus, when DNA from different individuals is cut with a restriction enzyme that cuts outside of the repetitive area, and is Southern blotted, using a probe that recognizes the repetitive sequence, patterns of DNA fragments are generated that are as unique as a person's traditional fingerprints. Hence the term, "DNA fingerprinting"!

Does gene therapy permanently "cure" genetic disorders?

In most cases, no. Because of the ethical constraints imposed upon gene therapy experiments, all of the human gene therapy experiments performed to date have involved *somatic cell gene therapy*. In other words, the tissues being modified via gene therapy are the somatic tissues (e.g., tissues making up the various non-reproductive organs and tissues of the body), not the reproductive cells or early embryos. Thus, any relief that is obtained is likely to be limited to the specific tissues targeted by the gene therapy. The patient may experience some improvement in health or relief of symptoms. However, they still possess the defective allele and can still transmit it to their children.

Key Terms

Site-specific mutagenesis

M13 phage

Embryonic stem cells

Totipotent

Blastocyst

Human genome project

Restriction fragment length polymorphism (RFLP)

Allele-specific oligonucleotide (ASO)

Sickle cell anemia

Variable number tandem repeats (VNTRs)

Ti plasmid

DNA fingerprinting

Yeast artificial chromosome (YAC)

Gene therapy

HIV virus

CD4 receptor

Severe combined immunodeficiency (SCID)

Tissue plasminogen activator (TPA)

Familial hypercholesterolemia

Low density lipoprotein (LDL) receptor

vegf (vascular endothelial growth factor)

Bovine somatotropin (BST)

For More Information . . .

Additional explanations on these topics are readily available:

For more information on site-specific mutagenesis, please see pp. 374-380 in Daniel Hartl and Elizabeth Jones, *Genetics: Principles and Analysis,* 4th edition. Sudbury, Massachusetts, Jones and Bartlett, Publishers, 1998. This topic is also discussed in depth on pp. 460-465 of Griffiths *et. al., An Introduction to Genetic Analysis,* 6th edition. New York, New York, W. H. Freeman and Co., 1996.

Want to learn more about RFLPs and VNTRs? These topics are covered on pp. 146-150 of Hartl and Jones, *Genetics: Principles and Analysis,* 4th edition. Sudbury, Massachusetts, Jones and Bartlett, Publishers, 1998. Additional information can also be found on pp. 482-486 and 509-511 of Griffiths *et. al., An Introduction to Genetic Analysis,* 6th edition. New York, New York, W. H. Freeman and Co., 1996.

An excellent discussion of gene therapy can be found on pp. 479-482 in Griffiths *et. al., An Introduction to Genetic Analysis,* 6th edition. New York, New York, W. H. Freeman and Co., 1996.

Want to read about Dolly, the first lamb to be cloned from somatic cells of an adult sheep? Then read the article by Wilmut *et. al.* (1997) in *Nature* 385: 810-813. 1997.

The Wise Owl Says

Where is the genetic engineering revolution leading us?

Are our children destined to grow up on a world in which assembly lines are operated by cloned factory workers? Where prenatal genetic testing will assure us a "perfect child"? Where we will purchase square, crush-resistant tomatoes at the supermarket to serve in a salad with our wilt-proof lettuce and our genetically-engineered, four-drumstick turkey ? Will we spend our weekends visiting the live dinosaur exhibition at the local zoo?

In his book, *Due Consideration: Controversy in the Age of Medical Miracles* , bioethicist Arthur Caplan states:

> "Do not fall for all the hype. Do not let those who learned about cloning from Woody Allen, Gregory Peck, Steven Spielberg and Michael Keaton frighten you into thinking that science and technology must inevitably be our master. Human beings can control the technologies they invent. To do so they must use their heads and not their genes."

Brave New World or Utopia? What do you think?

* Arthur Caplan. 1998. Due Consideration: Controversy in the Age of Medical Miracles. p. 35. John Wiley and Sons, New York.

Tough Nuts

In this "golden age" of biology, the power of recombinant DNA technology seems limitless. Genetic engineering discoveries will increase and practical applications will proliferate.

What will limit progress in this field? At present, it appears that technology will not limit progress. Instead, it is likely that social, economic, and political and civil progress will lag behind the technological developments. That is, a gap will probably emerge between science and society. Examples of these problematic areas are briefly in the section entitled "Social and Ethical Issues" at the end of Chapter 16.

Far Out

According to recent newspaper headlines..............

Differences in HIV Strains
May Underlie Disease Patterns

Potential for Use of Genetically Engineered Pathogenic Organisms
as Terrorist Weapons Grows

Naturally Immune HIV Patients May
Hold Key to Defense

Juries Avoid Dealing with
DNA Fingerprint Evidence

Journalists play a major role in shaping public opinion regarding scientific matters. Suppose you were a journalist........... Where do you think medicine and genetic engineering are headed?

What type of news article would you write, based upon one or more of these headlines?

Study Questions

1. How are "specific" mutagenic alterations introduced into a gene during site-specific mutagenesis experiments?

2. What is the advantage of using a phage such as M13 in site specific mutagenesis experiments?

3. Why are embryonic stem cells (ES) used when creating transgenic animals ("mammals made-to-order")?

4. What is the role of the *neo*^{r} gene in the selection for ES cells that have undergone successful transformation?

5. What new information will the Human Genome Project provide to scientists?

6. What are RFLPs? How do they arise?

7. How can RFLPs be used to diagnose genetic disorders such as sickle cell anemia?

8. What are VNTRs? How are they detected?

9. Could RFLPs associated with genetic disorders be used for purposes of DNA fingerprinting? Why or why not?

10. What is the advantage of using animals such as sheep to produce pharmaceutically-important products for therapeutic purposes?

11. What are some of the problems and challenges associated with gene therapy?

12. Outline the procedure first employed in gene therapy for severe combined immunodeficiency (SCID) in humans.

Answers to Chapter 16 Study Questions

1. Using an automated DNA synthesizer, we can synthesize a short (15-25 base-long) oligonucleotide which is complementary to the gene we want to alter, except that the synthetic oligonucleotide contains the altered base sequence. The synthetic oligonucleotide is allowed to anneal with a single-stranded copy of the normal version of the gene which has been cloned into an M13 cloning vector. Using the synthetic oligonucleotide as a primer, a complementary strand of the recombinant phage chromosome is synthesized *in vitro.* This double-stranded recombinant M13 phage DNA molecule can be used to transform *E. coli* cells and produce more copies of the *single stranded* , phage DNA molecule containing the *mutant* allele of our gene of interest.

2. M13 is a single-stranded (+) DNA phage that uses its (+) DNA strand to synthesize a complementary (-) DNA strand prior to DNA replication. The (-) strand serves as a template for the synthesis of multiple copies of the (+) stranded genome. If we introduce the genetically-altered allele into the (-) strand of the M13 cloning vector, multiple, single-stranded copies of our mutant gene will be produced during phage DNA replication.

3. Embryonic stem cells are undifferentiated and have the ability to be incorporated into any and all tissues in the developing embryo (e.g., they are *totipotent*). They can be removed from a blastocyst, cultured *in vitro,* undergo transformation with a genetically-engineered gene, and the transformed cells introduced into another blastocyst. In this new blastocyst they will become integrated into all of the tissues of the developing embryo, including the germline cells. Consequently, the chimeric transgenic animals can transmit their *new* gene to future generations of progeny via normal reproduction.

4. Successful transformation of ES cells involves homologous recombination between the "new" gene and the normal, homologous version of the gene already present in the genome of the ES cell. The recombinant gene has a promoterless *neo*^{r} gene linked to it, and is under the control of the recombinant gene's promoter. Neomycin resistance will only be expressed in ES cells that have successfully incorporated the *neo*^{r} gene into their genome adjacent to a functional promoter— hopefully the promoter of the "new" gene. ES cells are cultured in medium containing G418, a neomycin-like substance that kills mammalian cells that are not neomycin resistant.

5. The Human Genome Project will enable scientists to construct an accurate maps of the human chromosomes, including the locations of genes relative to each other and the locations of other landmarks, such as restriction sites, relative to the genes. Scientists will be able to determine the locations of genes associated with human genetic disorders, sequence these genes, and study their regulation and expression. Ultimately, it may become possible to offer treatment, diagnosis, and gene therapy for many human genetic disorders.

6. RFLPs **(restriction fragment length polymorphisms)** are variations in the length of DNA fragments generated when a DNA molecule is cut by a specific restriction enzyme. RFLPs arise as a result of mutations that abolish or create restriction enzyme cutting sites.

7. The mutation associated with sickle cell anemia abolishes or creates a cutting site for a specific restriction enzyme. Consequently, individuals who possess the mutant allele exhibit a *different* pattern of restriction fragments than normal individuals when their DNA is cut with a specific restriction enzyme and Southern blotted. Homozygous normal individuals will exhibit one pattern of band; individuals who are homozygous for the sickle cell allele will exhibit a second, different pattern of bands, and heterozygous individuals will exhibit both the normal and sickle cell patterns of bands.

8. VNTRs or **Variable Number Tandem Repeats**, are short base sequences that are tandemly-repeated and which occur at various locations throughout the genome. The number of copies of the repetitive sequence vary from location to location within the genome of a particular individual, and different individuals exhibit different patterns of VNTRs. VNTRs are detected by cutting the DNA of an individual with a restriction enzyme that cuts *adjacent to, but not within the repetitive sequence.* The DNA is then Southern blotted, using a probe that hybridizes with the repeat sequence. A series of bands (DNA fingerprints) are generated which correspond to clusters of repetitive sequences containing different numbers of repeats.

9. No. There are only a few different alleles of a gene within a population, so you wouldn't expect to find much variation between individuals with respect to RFLP patterns correspond to a specific gene. VNTRs exhibit tremendous variation among individuals, both in the number of locations at which the repetitive sequence occurs within the genome, and the number of copies of the repeat at each location.

10. First, since sheep are eukaryotes, one does not have to deal with the problems associated with the expression of eukaryotic genes in prokaryotes, namely the inability of prokaryotic cells to recognize eukaryotic promoters, remove introns, splice exons together, translate eukaryotic mRNA on prokaryotic ribosomes, and properly process eukaryotic proteins. Second, the gene can be engineered to be expressed only in mammary tissues, so that the protein will be excreted in the milk. Milk is clean, readily accessible, can be obtained without harm to the experimental animal, and it is possible to isolate and purify large quantities of the desired transgenic protein from it.

11. Before gene therapy can be attempted, the gene in question must be identified, cloned and sequenced. An appropriate vector must be available to deliver the gene to the affected tissue. The tissue must be accessible to the vector. In some cases, the cells can be removed from the body and genetically modified; in other cases, the tissues cannot be cultured or modified outside of the body, so ways must be developed to deliver the functional gene to the affected tissues. The site at which the new gene is inserted into the genome can also cause problems. Under ideal circumstances, the functional gene would undergo homologous recombination with the defective gene in the host. In many instances, the new gene is randomly inserted into the genome, and may actually *cause* problems by inactivating a normal, non-homologous gene into which it is inserted.

12. T lymphocytes were removed from the patient and mixed with a genetically-engineered retrovirus containing a normal copy of the **adenosine deaminase gene**, the gene that is defective in SCID victims. The modified T cells were tested to verify that they contained a copy of the ADA gene, and that the gene was functioning normally. The T cells were transfused back into the same patient, where the functional ADA gene produced an elevation in her serum ADA levels and a rise in the white blood cell count.

Concept Map

A concept map may help us to better understand the applications of genetic engineering!

Site-Specific Mutagenesis

Diagnosis of Human Genetic Disorders

Human Genome Project

Gene Therapy

Transgenic Animals

DNA Fingerprinting

Please include the following terms in your concept map:

RFLPs	G418 medium	Yeast artificial chromosomes
VNTRs	ES cells	Adenosine deaminase gene
ASO probe	Ti plasmid	Severe combined immunodeficiency
M13 phage		

P.S.: Please add additional terms from Chapter 16!

Science and Society Issues

In the early 1970s, when gene cloning experiments were being carried out for the first time, practical issues associated with health and safety emerged. These issues were vigorously debated in newspaper columns, city councils, and corporate headquarters. Many scientists were worried that *E. coli*, the usual host for plasmid and viral vectors, might infect a human being while carrying a potentially harmful vector. After all, it was reasoned, *E. coli* is a normal inhabitant of the human intestinal tract. Of special concern was the possibility that some of the vectors might carry cancer-causing genes that could infect and kill human hosts. Another concern was the possibility that new strains of host bacteria, if permitted to spread outside the laboratory, might wreak havoc with the natural environment.

In 1975, a conference of scientists decided to establish guidelines for "recombinant DNA" experimentation. With the direction of the US National Institutes of Health, a set of regulations was adopted. Included were rules that mandated the use of genetically handicapped bacterial hosts that cannot multiply when released to the natural environment. In addition, rules concerning the design of special isolation booths and ventilation systems were implemented, so that laboratory procedures would be as safe as possible.

Eventually, it became apparent that many of the original fears associated with recombinant DNA safety were exaggerated, and more relaxed guidelines were adopted. Presently, most recombinant DNA procedures are carried out on the bench top, without taking special precautions such as using isolation booths, positive ventilation systems, or air filters. Only during the study of genes associated with pathogenic diseases, such as some viral infections, are extensive containment facilities employed. The public debate over safety issues has largely subsided.

New issues associated with recombinant DNA technology have, however, emerged. These debates often revolve around social, economic, moral, and ethical issues, rather than technical matters, such as safety precautions. Literally, dozens of debates have been initiated concerning applications of recombinant DNA technology. A brief outline of several of the major issues and their associated questions follows:

Farm animals:

A growth hormone has been produced by genetic engineers. This protein has the capacity to improve growth rates and increase milk production of dairy cows. With increased milk production, fewer dairy cows will be required to supply the demand for milk products. Small, family-oriented farms account for the majority of milk production. Should they be put out of business by more efficient farms that use growth-hormone-stimulated dairy cattle?

Molecular medicine:

Tissue plasminogen activator (tPA) is a protease (an enzyme that degrades proteins) that activates the human body's own plasmid protein. Plasmin is a protease that cleaves the major blood clot protein, fibrin. If tPA is given to the victim of a heart attack soon after the attack, the blood clots that caused the coronary blockage will be dissolved and the victim's health will not be further jeopardized. In order to recover the costs associated with the research, development, and marketing of tPA, and to generate corporate profits, a high price is charged for it by the pharmaceutical company that produces it. Many potential heart attack victims cannot afford this high price. Should a heart attack victim's chances of surviving permanent damage be directly related to the victim's ability to pay for such expensive treatments?

Environmental hazards:

Various desirable traits can be moved from one plant to another. For example, protein content, high in some plants, can be transferred into other, less well-endowed plants. Likewise, traits such as cold resistance can be taken from one species, such as fish, and engineered into crop plants. What might be the effect on the ecology of a well-established environment in the event of an inadvertent release of these novel plants?

Genetic disease diagnosis:

As the list of genetic diseases that can be identified in tissue samples increases, and as the costs for performing such tests decreases, health and life-insurance companies will want to employ genetic diagnoses for identifying high-risk applicants. Should insurance companies be permitted to reject applicants based solely on genetic test results?

Human genome sequencing:

Within a decade, the nucleotide sequence of the entire human genome will be known. All three billion base pairs will be ordered as a linear DNA sequence! Should an individual's right to privacy be maintained once it becomes possible to compare the nucleotide sequence of one person with that of another person?

These and other related issues have generated much discussion among scientists as well as nonscientists. In fact, now more than ever, the moral, social, economic, and ethical implications of modern science are being examined.

APPENDICES

APPENDIX A: Chemical Principles for Molecular Biology

A brief word about chemical bonds

An atom is made up of a nucleus and its electron shells. Inside the nucleus there are protons with positive charges and neutrons with no charge. The electron shells are made up of the negatively charged electrons that orbit the nucleus. The charges on the positive protons and the negative electrons serve to cancel each other.

Atoms bond together to form molecules of specific shapes. The molecule's shape is dependent on the bonds it makes. Like pieces in a puzzle, the parts of the molecule must fit correctly. If the bonds are disturbed and a piece changes shape and no longer fits, then the entire puzzle is left incomplete, or, in the case of a macromolecule, nonfunctional.

Let's review bond types, shall we?

I. COVALENT BONDS (sharing)

Atoms that have completely filled outer shells are stable and do not readily react with other atoms. However, those atoms with incomplete or missing electrons are unstable and therefore "seek out" electrons to stabilize themselves.

Most of the unstable atoms, especially the smaller ones (H, C, O, N, S, P), can be found in biological compounds. By bonding together they form stable structures. They can share more than one electron. The number of electrons needed to fill an atom's outermost shell determines the number of bonds that the atom can accommodate. Also, the smaller the atom the stronger the bond.

Question #1: In molecular biology, what do we call the *structure* that is formed by covalent bonding? See Chapter 4 in *EMB* for the answer.

II. IONIC BONDS (transferred)

When an electron is transferred from one atom to another atom, stabilizing both of them in the process, it forms an ionic bond. Once one atom has lost a negative charge and the other has gained it, they are left with opposite charges and attract like magnets. Please see Figures 2-10 and 4-6 in *EMB* for examples of ionic bonds.

III. POLAR COVALENT BONDS (partial)

This bond is similar to a covalent bond in that electrons are shared. One atom, however, appears to be selfish. It is usually stronger and holds its electrons tighter, so that the other atom gets only enough of the charge to make the molecule stable. The molecule is, therefore, left with one end more negatively or positively charged than the other.

Question #2: What is a good example of a polar covalent molecule? Please see Figure A-7 in *EMB* for the answer.

IV. HYDROGEN BONDS

Hydrogen is positively charged and seeks out negative charges. It often aids in forming other polar covalent bonds. Although they are weak bonds, multiple hydrogen bonds can together be very strong. They are very important in the structure of proteins and nucleic acids. They can be responsible for holding a single molecule in its shape (in*tra*molecularly), or for holding one or more molecules together (in*ter*molecularly).

Question #3: Hydrogen bonds are *very* important for holding together which large macromolecule? Please see Figure 3-2 in *EMB* for the answer.

V. HYDROPHOBIC (water-hating)

Nonpolar compounds have no partial charge. That is, they are neutral. They cannot, therefore, bond with the positively charged hydrogen atoms in water. To avoid the charge of water (H_3O^+), hydrophobic molecules clump or bond together.

VI. RESONANCE BONDS

The concept of resonance is necessary because of the limitations in the way *we write* structures. There may be more than one correct way to write a molecular structure with resonance. For example, ozone can be written:

O-O=O or O=O-O

However, through experimentation it is known that each bond is functionally equal to the other. One is not a single bond and the other is not double. There is another bond with a strength and length which is in between the two so the actual molecule is always the same. Sometimes resonance is shown with a dashed line.

VII. PEPTIDE BONDS

Peptide bonds link amino acids together to form a polypeptide or a protein. A peptide bond is formed when a carboxyl group bonds to an amino group. When the OH and the H react (due to the action of an enzyme) they form H_2O (water), which is released.

This reaction is also diagrammed in Figure 2-3 of *EMB*.

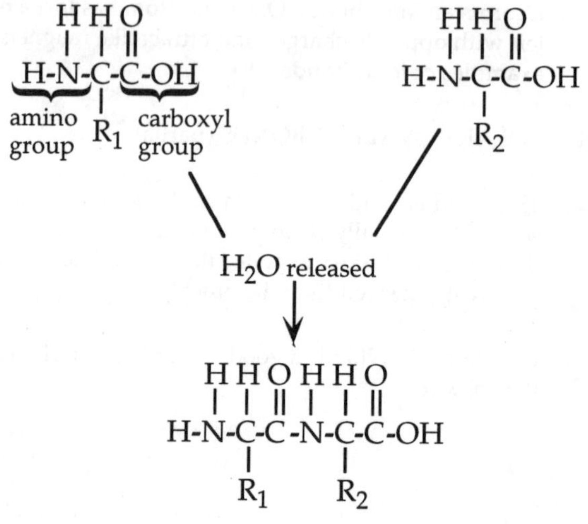

Answers to the questions:
1. primary structure
2. water
3. double-stranded DNA

A brief word about *organic compounds*:

All organic molecules are structured around the carbon atom. Carbon is an extremely versatile element, and the fact that it can form strong bonds to hydrogen, oxygen, nitrogen, and sulfur makes it important for the optimal functioning of biological systems.

Let's have a review of hydrocarbons:

There are many classifications of molecules depending on their composition. A major branch of organic chemistry deals with hydrocarbons. Hydrocarbons are molecules that contain only carbon and hydrogen, and can be divided into 2 groups: **aliphatic hydrocarbons** and **aromatic hydrocarbons**. (When you think of hydrocarbons, think of motor oil, gasoline, and butter among others.) Aliphatic hydrocarbons can be divided further into 3 more groups:

1. **Alkanes** are molecules that follow the general formula C_nH_{2n+2} (where **C**=carbon and **H**=hydrogen). Thus, a molecule that contains 2 carbon atoms would contain 6 hydrogens. Alkanes are said to be saturated hydrocarbons. This means that every carbon contains as many hydrogens as the octet electron pairing rule will allow (i.e., every carbon only contains single bonds).

   Problem #1: What is the formula for the following compound? What is its name?

   $$CH_3–CH_2–CH_2–CH_2–CH_2–CH_3$$

2. **Alkenes** are aliphatic hydrocarbons that are not saturated. This means that one or more carbons contain carbon–carbon double bonds. The double bonds of alkenes are extremely reactive, making alkenes extremely important in industrial processes (i.e., the polymerization of plastics). Alkenes are named like their cousins, the alkanes. The only difference is that the "-ane" ending is replaced with "-ene."

   Examples:

$CH_3–CH_2–CH_2–CH_3$	$CH_3–CH=CH–CH_3$
Butane (alkane)	Butene (alkene)

   Pages Figures A-15 and A-16 in *EMB* illustrate several more examples.

3. **Alkynes** are molecules that contain carbon–carbon triple bonds. We will not be concerned with them here, for they are not frequently encountered in biological systems.

   Isomerism is a characteristic of many organic molecules. Structural isomers are molecules that have the same molecular formula, like C_2H_6, but have a different orientation of bonds. Geometric isomers are molecules that have the same orientation of bonds, but have some groups arranged differently in space.

Example:

Structural | **Geometric**

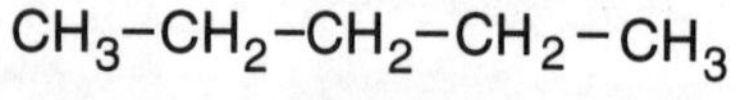

Pentane

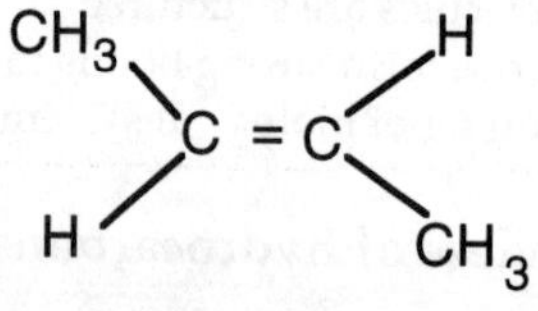

Trans-2-butene

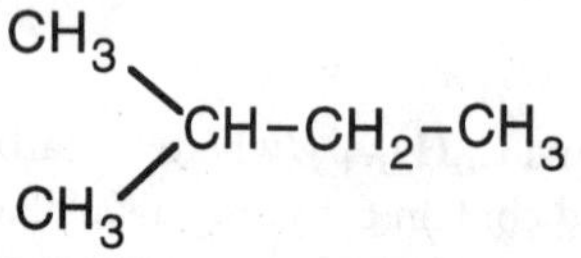

Isopentane

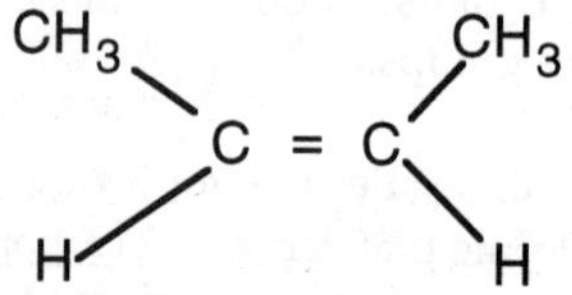

Cis-2-butene

Hydrocarbons also occur in the form of cyclic (aromatic) compounds. **Cycloalkanes** are alkanes that are connected in a ring and follow the formula C_nH_{2n}. They are, however, still saturated. **Cycloalkenes** are cyclic molecules that are not saturated because they contain one or more double bonds. Here are some of their shapes:

Cyclohexane (Cycloalkane)

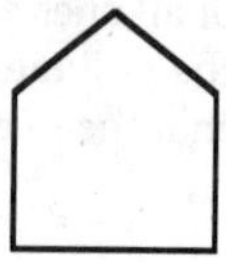

Cyclopetane (Cycloalkane)

Cyclohexene (Cycloalkene)

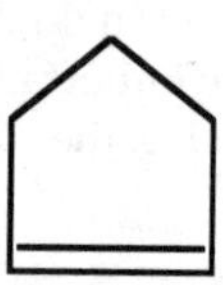

Cyclopentene (Cycloalkene)

Problem #2: Is the following compound cyclic or linear? What is the name of the compound and what does it look like?

$$C_5H_{10}$$

Here's one more type of carbon compound we need to become acquainted with:

Carboxylic Acids are a group of molecules that are very important in biological systems. They contain a carbonyl group which looks like this:

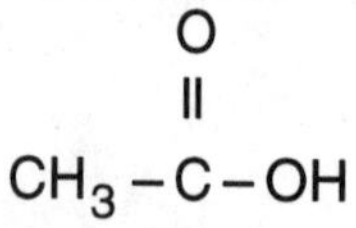

Carboxylic Acid

(A carbonyl group contains a carbon bound to an oxygen in a double bond)

Carboxylic acids are important in the formation of the peptide bond discussed above. The electrons in the double bond are in resonance, and the resonance structures look like this:

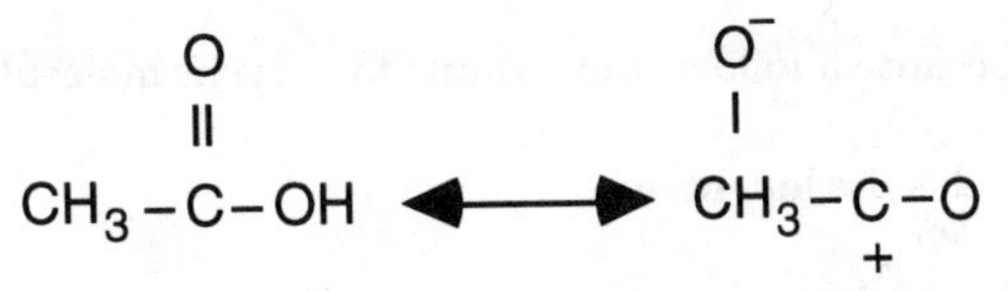

Carboxylic Acid Resonance Structures

As you can see there now exists a partial positive charge on the carbon atom bound to the oxygen. This partial positive charge is important because many reactions depend on its presence.

Lastly, let's consider nitrogen compounds:

Compounds containing nitrogen are extremely important in biological systems. These compounds come in the form of **amines, imines,** and **amides.** Please see the Appendix in *EMB* for descriptions of the structure of the nitrogen-containing compounds.

A very important reaction in biological systems is the polymerization of amino acids to form proteins discussed above. The reaction occurs between the amino group of an amino acid and the carboxylic acid group of another amino acid.

Problem #3: Based on the following clues and the knowledge you have gained, try writing a reasonable explanation for the reaction which would occur. Illustrate the reaction with a diagram.

1. Carbon desperately wants its eight valence electrons.
2. Nitrogen prefers to have only three bonds but can form 4 bonds for a brief period of time.
3. The –OH group of the carboxylic acid can borrow a proton from the amino group.
4. H_2O can be removed from the amino acid easily.
5. The electrons in the double bond are free to move onto the oxygen at any time.

Answers to Problems

Problem #1

The formula for the compound is: $\mathbf{C_6H_{14}}$
The name of the compound is: **n–Hexane**

Problem #2

The compound is **cyclic** because it follows the formula for a cyclic molecule:
$\mathbf{C_nH_{2n}}$
The name of the compound is **Cyclopentane**
It looks like this:

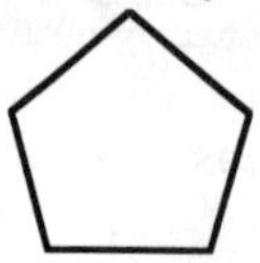

Problem #3

One of the most important aspects is the fact that there exists a partial positive charge on the carbon of the carboxylic acid (look at the resonance diagram). The nitrogen of the amino group, being a good nucleophile, can form a bond with the carbon. Now the carbon has eight valence electrons. The nitrogen has 4 bonds now, but this problem will be taken care of. The –OH group of the acid, which is basic, can borrow one of the protons from the nitrogen. Now the nitrogen has its 3 bonds, and you have also formed a water molecule. Now, at random, the electrons that were on the oxygen can move back down to reform the double bond with carbon. In the process, since carbon can only have 4 bonds, the water molecule unbinds the carbon and it goes its separate way. Now you have just formed a peptide bond!

APPENDIX B: Instructor's Guide

Introductory comment

This study guide is meant to serve as a key component in a comprehensive teaching program for an introductory level molecular biology course.

The focus of the course taught at Indiana University is on "learning," rather than "teaching." Accordingly, a wide variety of learning aids are employed. They are briefly described on the following pages. A formal "collaborative learning" approach is taken in which the professor relinquishes authority and control over information transfer. Instead, this aspect of the course is left to the textbook (*EMB*), while the professor emphasizes the development of critical thinking (i.e., problem-solving) skills with the aid of this Student Manual.

The format of the textbook is adhered to very closely. Approximately one chapter is covered each week of the 15-week course (three 50-minute lectures per week). Rather than "recite" the textbook during lecture, students are advised to read and study the textbook on their own time. Instead, lecture time is used to explain in detail only the most important and/or difficult portions of each chapter. Students may then supplement the explanations given in lecture by reading the "Here's Help" sections in this Student Manual.

Background information and supplementary material for teaching a sophomore-junior level molecular biology course can be found in the following books:

Encyclopedia-type textbooks:

I. Alberts, B., Bray, D., Lewis, J., Ruff, M., Roberts, K., and Watson, J.D. 1994. *Molecular Biology of the Cell*, 3rd edition. (New York: Garland Publishing, Inc. 1294 pages. ISBN 0-8153-1619-4 (hardback), ISBN 0-8153-1620-8 (paperback)).

A splendid introductory text that covers some of the same subjects as *EMB*. Although coverage of molecular biology tends to be light, this 1,294 page textbook encompasses virtually all aspects of modern biology; from evolutionary considerations to descriptions of the human nervous system. The illustrations are usually good and can occasionally be used to supplement the even simpler illustrations found in *EMB*.

II. Watson, J.D., Hopkins, N.H., Roberts, J.W., Steitz, J.A., and Weiner, A.M. 1987. *Molecular Biology of the Gene*, 4th edition. (Menlo Park, CA: Benjamin/Cummings Publishing Co., ISBN 0-8053-9612-8)

A detail-oriented, advanced textbook that covers virtually all of the subjects included in *EMB* at a more sophisticated level. This textbook contains nice illustrations, photographs of experimental data (e.g., gels), and dozens of tables of factual information. Although too advanced, and much too pedantic, for an introductory course in molecular biology, this is an excellent reference book.

III. Lewin, B. 1990. *Genes VI*. (New York: Oxford University Press. ISBN 0-19-85777-8)

An authoritative, advanced textbook that is most suitable for first-year graduate courses and as a reference book for instructors of undergraduate courses. The illustrations are simpler than those found in the Alberts et. al. and Watson et. al., thus making them easier to understand. Valuable insights into research topics are provided that may aid instructors in projecting the possible future directions of various subdisciplines in molecular biology.

IV. Maloy, S. R., Cronan, J. E., Jr., and Freifelder, D. 1994. *Microbial Genetics*, 2nd edition. (Boston: Jones and Bartlett Publishers, ISBN 0-86720-248-3)

Contains more detailed treatments of many of the topics included in *EMB*. The illustrations are the simple "stick diagram" type featured in *EMB*. In fact, some of the same illustrations are used in both books, but this textbook expands the treatment of the topics which overlap.

Workbooks:

V. Watson, J.D., Gilman, M., Witkowski, J., and Zoller, M. 1992. *Recombinant DNA*, 2nd edition. (New York: Scientific American Books (W.H. Freeman and Co.) ISBN-0-7167-2282-8 (paperback))

An excellent description of current research themes that are driven by recombinant DNA technology. The illustrations are multicolored and well-designed. Although too advanced for undergraduate courses, this makes an excellent reference manual for instructors.

VI. Alcamo, I.E.. 1996. *DNA Technology: The Awesome Skill.* (Wm. C. Brown Pub.., IA. ISBN 0-697-21248-3.)

A useful study guide which is well illustrated and especially user-friendly (for undergraduate students).

VII. Glick, B. R., and Pasternak, J. J. *Molecular Biotechnology*. 1994. (Washington, D. C.: ASM Press, ISBN 1-55581-071-3)

Written at a junior/senior undergraduate level, this text focuses on both pure and applied aspects of genetic engineering. Special attention is given to plants and transgenic animals.

Molecular Biology is taught at the front-end of the Indiana University curriculum.

As Molecular Biology emerged as a discipline, it was initially incorporated in the curriculum either at select intervals, or tacked onto the tail-end of several courses, so that students would be informed of the latest developments in the field. Then, as Molecular Biology came to be regarded as a separate discipline, it was taught as a first-year graduate course, open only to select senior-level students. Still later, it became a senior-level undergraduate course.

Approximately eight years ago, a faculty committee reviewed the undergraduate biology program and repositioned Molecular Biology to the front end of the curriculum. The rationale behind this was multifaceted, but emphasized the impact of Molecular Biology on cell biology, developmental biology, genetics, and even evolutionary biology. This impact was so dramatic and overpowering that it deserved positioning as an entry-level course for undergraduate biology majors. Then, subsequent courses could presuppose fundamental knowledge of Molecular Biology so that excessive repetition of relevant molecular concepts and techniques could be avoided.

Initially, many faculty claimed that employing Molecular Biology as a starter course would be counterproductive. It was feared that most sophomore students would be unable to cope with a discipline that emphasized mechanisms, biochemical methods, and experimentation. In other words, they were apprehensive that a starter course in Molecular Biology would be ineffective because it would need to be so watered down and descriptive.

Those fears were quickly allayed when the decision was made to use a brief text (the original edition of *Essentials of Molecular Biology*) and to recruit some good instructors for the course. Presently, Molecular Biology, a one-semester (15-week) course, is followed by Genetics and Development, Cell Biology, and Evolution-Ecology. The Department of Biology used the impetus provided by a generous multiyear Howard Hughes Medical Institute Undergraduate Initiative grant to build "critical thinking" into the curriculum. Revision of *Essentials of Molecular Biology* and the development of this student-oriented study guide represent the products of this effort to achieve a more intellectually complete curriculum. It is expected that additional efforts will ultimately result in a complete undergraduate biology critical-thinking curriculum.

Based upon EMB, "Cooperative learning" approaches can be employed to enhance student achievement levels in molecular biology courses. Here is how we organize a cooperative learning format at Indiana University:

Lectures: Three 50-minute lectures are presented each week by the professor. The lectures follow the textbook (*EMB*) closely. Virtually all illustrations, figures, and tables from *EMB* are converted to "overhead projections" for use in lecture.

Rationale:	Students need not take extensive lecture notes. Instead, they can focus on obtaining an understanding of each phenomenon. As well, students can read ahead and plan their schedules around the pace provided by the textbook.

Learning group (LG) meetings: Once per week students meet in fixed groups of six to complete a worksheet that covers the previous week's lecture presentations. Each learning group is guided by an "undergraduate teaching intern." The role of the LG leader is to assist and encourage peer group learning experiences.

Rationale:	Worksheets* bring to life the concepts, facts, and descriptions presented in *EMB*. Each worksheet is comprised of a list of problems and conceptual questions. This design enables students to develop "critical thinking" (i.e., analytical) skills during the course of their semester of molecular biology.

Study Hall: Once per week an "open" discussion meeting is held by the professor. At this time, relevant material is reviewed. In addition, student inquiries pertaining to the material are addressed. Students are encouraged to use this manual for clarifying and expanding information presented in *EMB*.

Rationale:	Small-group learning situations are invariably more productive than individual study or passive experiences such as attendance at lectures. As well, LG leaders encourage students to use their verbal communication skills so they feel less intimidated about asking questions.

Learning aids: Each student is provided a free kit$^{\Psi}$ for aid in understanding replication, transcription, and translation. The kits contain the appropriate pipe cleaners, beads, and labels for preparing three-dimensional models of DNA undergoing these three processes. These kits are assembled during lecture, study hall, and in learning groups.

Rationale:	Students display a variety of learning styles. While some students excel at converting two-dimensional illustrations into three- (or four-) dimensional representations, others find it very difficult. Simple pipe-cleaner models provide a convenient and inexpensive learning tool for beginning students of molecular biology.

Class size: Each semester (i.e., three times per year) at Indiana University, L211 (Molecular Biology) is taught as a 3-credit hour course. It enrolls approximately 300 students each semester. Nevertheless, since learning groups are used, the class remains "functionally" small.

*A sample worksheet is included at the end of this manual. Instructors may request a complete set of worksheets by writing to George M. Malacinski, Ph.D.; Professor of Biology, Department of Biology, Indiana University, Bloomington, IN 47405.

$^{\Psi}$Sample kits are available to instructors who adopt this owner's manual for general class use. Please contact George M. Malacinski, (email: malacins@indiana.edu).

A comprehensive, student-oriented learning program for molecular biology is achieved by combining all of the components of L211 (i.e., *EMB* text, this Student Manual, learning groups, pipe cleaner models, and study hall). Sophomore- and junior-level students respond very well. In fact, a comparison of student achievement levels from this comprehensive "cooperative learning" format with the traditional lecture-only, encyclopedic textbook approach to teaching molecular biology, reveals an astonishing gain in student learning when cooperative learning is employed.

The Indiana University Biology Department is developing a critical thinking agenda for the undergraduate curriculum.

This revision of the textbook *Essentials of Molecular Biology* and development of this study guide are directly linked to the Critical Thinking Agenda (CTA) being developed in Indiana University's Biology Department. The potential power of Molecular Biology, as both a technical and intellectual force, was early recognized by research scientists. The power of Molecular Biology for the development of critical thinking skills in undergraduates has, however, been more slowly acknowledged. At Indiana University, Molecular Biology has been introduced early in the curriculum for biology majors. As mentioned in the Prologue of *EMB*, the main driving force for repositioning Molecular Biology to the front end of the curriculum was its impact on contemporary cell, genetic, developmental, and evolutionary biology. Another driving force was the desire to use instruction in Molecular Biology to develop critical thinking skills early in the undergraduate experience.

"Critical thinking" represents the use of analytical skills and reasoning powers to understand specific phenomena and to synthesize relationships between different phenomena. These analytical and reasoning skills are developed through the CTA by focusing problem-solving skills and encouraging students to become aware of their own thought processes. The ultimate goal of the CTA is to develop intellectual dexterity in students. This is characterized by analytical thought, objectivity, a sense of fairness, and the ability to diagnose cause and effect relationships in order to make logical connections.

Compelling reasons exist for undertaking an ambitious program such as Indiana University's CTA:

1. The base knowledge of contemporary biology has expanded exponentially. It has now reached the point where analytical and reasoning skills are necessary to assess the relative merit, or importance, of new information. Old information does not usually become obsolete; nor is it usually proven wrong. Moreover, new information is not necessarily relevant. The amount of both old and new information, in virtually all fields of biology, has reached dizzying proportions. The role of the scholar is gradually shifting from one of accumulating information ("knowing") to one of "assessing," or "evaluating" information. Therefore, critical thinking skills are clearly necessary and should be developed in students as early as possible in their academic careers.

2. The changing nature of the workplace requires participants who are skilled learners, analyzers, and synthesizers. Attempts to teach students a body of knowledge that directly translates into job skills is, for biologists, like trying to hit a moving target. By focusing on intellectual development, the CTA gives biology undergraduates the flexibility that is being increasingly required in the workplace.

3. The demand is growing for professional biologists by commercial firms, government agencies, and academic institutions. The CTA deliberately takes a "user friendly" approach to Molecular Biology in order to encourage, rather than discourage, sophomore students to pursue a major in biology.

Indiana University is fortunate to have received generous financial support for the CTA from the Howard Hughes Medical Institute Undergraduate Initiative. The CTA approaches student development from the perspective advanced by Perry[1], Chickering[2], and others in that the acquisition of intellectual skills is linked to personal growth. Accordingly, the learning skills that comprise the CTA are arranged in a hierarchical fashion:

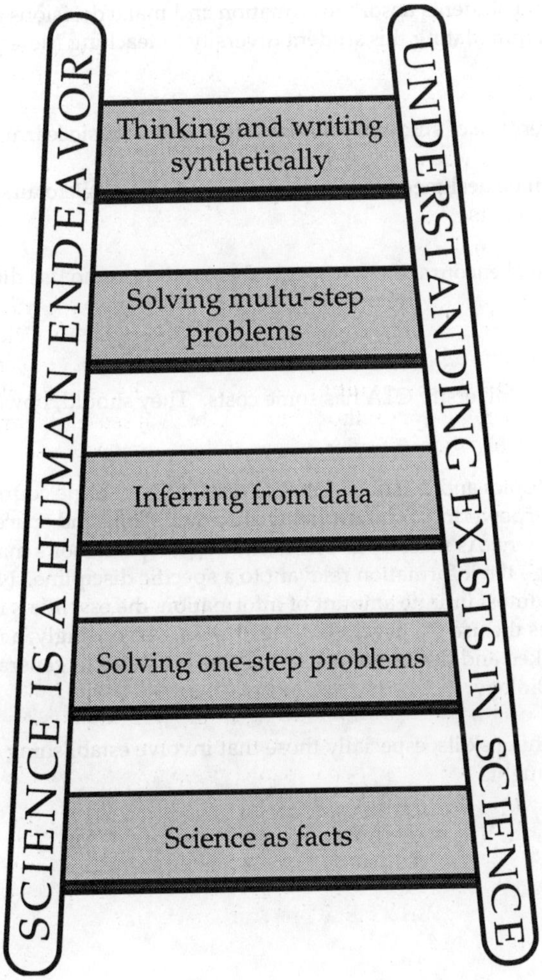

[1]Perry, W. *Forms of Intellectual and Ethical Development in the College Years*. (New York: Holt, Rinehart, and Winston , 1970).
[2]Chickering, A. *Education and Identity*. (San Francisco: Jossey-Bass, 1969).

Early courses emphasize science as representing a body of facts and a way of knowing. It is also emphasized as a human endeavor. Students are guided in their efforts to learn problem-solving skills. During the later years in the curriculum, students are encouraged to appreciate uncertainty as a fundamental feature of science. Their personal beliefs and thinking habits are challenged and their abilities to synthesize relationships between seemingly dissimilar information are perfected.

A wide range of personality traits are required by each student in order to progress up the Intellectual Skills Ladder. Since these skills are possessed in varying amounts by individual students, the Indiana University CTA has used personality profiling of students to develop appropriate teaching strategies. Personality profiling acknowledges that different personality types usually have different learning styles. Variation in the manner that students absorb information and make decisions results in different problem-solving strategies. By accommodating this student diversity in teaching Molecular Biology, several benefits accrue:

1. A higher percentage of students will develop stronger basic learning skills.

2. All courses can be designed to be "user-friendly." No single course (e.g., Genetics) need emerge as a "weed-out course."

3. Students will feel encouraged to respect other students who use different problem-solving strategies.

Employing the Indiana University CTA has some costs. They should, however, not be considered prohibitive:

First, the range of topics and extent of detail covered in any single course must be carefully monitored in order to avoid overpowering or intimidating students. Time and space must also be left for intellectual development. As recently as one decade ago, a professor's main task was to collect, organize, and package the information relevant to a specific discipline. Now, however, the main task is to sort out, from an almost infinite amount of information, the essentials for a particular discipline. This consideration negates the use of encyclopedic textbooks. Accordingly, a revision of the original edition of *EMB* was undertaken and this Student Manual was prepared to generate a user-friendly approach to learning Molecular Biology.

Second, basic reasoning skills, especially those that involve establishing cause and effect relationships, need to be actively taught.

Chapter 3 Worksheet

1. This question refers to this molecule:

   5′ ATCATCGGTAGTCATGTACCGATGAA 3′
   3′ TAGTAGCCATCAGTACATGGCTACTT 5′

   a. The underlined portions of this DNA sequence are palindromic. What possible conformation could this molecule assume other than the linear double-strand shown here?

   b. Draw a schematic of what this structure would look like. (Exclude the actual base sequence; just make a quick sketch.).

   c. Under what conditions would this structure be likely to form? How might this be advantageous to the stability and/or function of the DNA molecule?

2. You have a DNA solution about which you know very little. In order to better characterize the solution, you decide to do denaturation and renaturation analyses.

   a. What is denaturation and what is renaturation?

   b. What sort of information is gained by denaturation and by renaturation?

   c. We first decide to observe the solution's absorption of UV light as we raise the temperature. We note that the A_{260} rises and then plateaus at a level of 1.37. Describe (in one sentence) the physical state of the molecule at this point.

d. We now let the solution cool, but before doing so, we lower the salt concentration to 0.01M and again observe the value of A_{260} . By allowing the solution to cool, we would expect A260 to drop as well, but for some reason A_{260} still equals 1.37. What is happening in this scenario?

e. Noticing that something must be wrong with our experimental setup, we adjust the conditions such that [NaCl] is 0.3M and the temperature is 25°C below the T_m we measured for the solution. Now conditions are ideal for renaturation. Your sample was obtained by doing PCR on a short (70 b.p.) sequence of DNA. Thus you have many copies of a short DNA sequence. Ignoring actual values, predict and draw what a renaturation curve would look like for our sample. (Remember to label the axes.)

3. While denaturing a highly repetitive DNA sample, you accidentally contaminated it (foreign pieces of DNA have fallen into the Eppendorf tube). You need to recover your denatured sample before you can continue with your analysis. If you are provided with a very small portion of your original DNA sample, how could you recover your target DNA from the contaminated Eppendorf? Explain or diagram your procedures in detail.

4. Regarding DNA sequencing:

a. Name two methods used to determine a DNA sequence and briefly explain each.

b. Draw and explain the difference between ribose, deoxyribose, and dideoxyribose sugars.

c. What happens when a dideoxynucleotide is incorporated in the growing strand?

d. What enzyme facilitates the construction of polynucleotides?

e. What other reagents are required for the Sanger reaction to occur?

f. Why is the 5′ → 3′ direction of the sequence read from the bottom of the gel upwards?

g. What is the electrical polarity ("+" or "-") of the bottom of the gel and why?

h. What would be the original DNA sequence from the following gel (don't forget polarities):

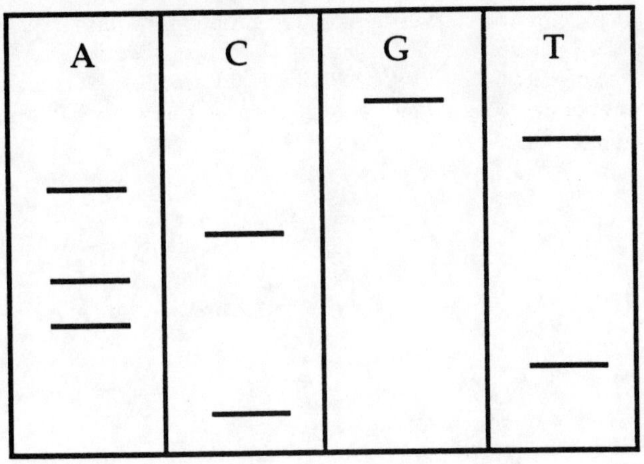

Gel reads:

Original DNA sequence:

17. Regarding DNA sequencing

a. Name two methods used to determine a DNA sequence and briefly explain each.

b. Draw and explain [illegible]

c. [illegible]

d. W[illegible] polymeriz[illegible]

e. What other components are required for the Sanger reaction to work?

f. Why is the [illegible] of the [illegible] bottom [illegible]

g. What [illegible] gel and why?

h. [illegible] from the following [illegible]

Original DNA sequence: